T0269296

The chemistry of macrocyclic ligand complexes

The chemistry of macrocyclic ligand complexes

LEONARD F. LINDOY

Department of Chemistry and Biochemistry,
James Cook University, Townsville, Australia

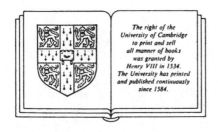

The right of the
University of Cambridge
to print and sell
all manner of books
was granted by
Henry VIII in 1534.
The University has printed
and published continuously
since 1584.

CAMBRIDGE UNIVERSITY PRESS

Cambridge

New York   Port Chester

Melbourne   Sydney

Published by the Press Syndicate of the University of Cambridge
The Pitt Building, Trumpington Street, Cambridge CB2 1RP
40 West 20th Street, New York, NY 10011, USA
10 Stamford Road, Oakleigh, Melbourne 3166, Australia

First published 1989
First paperback edition 1990

British Library cataloguing in publication data
Lindoy, Leonard F.
   The chemistry of macrocyclic ligand complexes.
   1. Macrocyclic ligands
   I. Title
   547'.5

Library of Congress cataloguing in publication data
Lindoy, Leonard F.
   The chemistry of macrocyclic ligand complexes/Leonard F. Lindoy.
      p.   cm.
   Bibliography: p.
   Includes index.
   1. Ligands.   2. Cyclic compounds.   I. Title.
QD474.L54 1989
541.2'242–dc19   88-1096

ISBN 0 521 25261 X hardback
ISBN 0 521 40985 3 paperback

Transferred to digital printing 2003

PN

Contents

Preface

The metal-ion and host-guest chemistry of macrocyclic ligands has developed rapidly over recent years and now impinges on wide areas of both chemistry and biochemistry. It is the aim of this work to present, in a single volume, an overview of the main developments in the area. An attempt has been made to provide a text for those seeking an introduction to the field as well as one which will be of use to practising 'macrocyclists'. In particular, it is hoped that the material covered will provide an appropriate basis for a senior undergraduate/graduate course on macrocyclic chemistry. It has been used in this way by the author at James Cook University.

The text should also serve to put in a chemical perspective the several types of macrocyclic systems found in nature – as such, it should be of benefit to those whose principal interests are the biochemical aspects of such cyclic systems.

Throughout the text, an effort has been made to include sufficient references to provide a balanced introduction to the relevant literature. This has not always been straightforward since frequently a number of excellent papers were available to illustrate a particular point. When this occurred the decision of what to include was somewhat arbitrary. I apologize to the many authors whose work it has not been possible to cite.

It is a particular pleasure to acknowledge my many friends and colleagues, on five continents, who have contributed in one way or another to this work. They are too numerous to name individually. However, I wish to express my special gratitude to Professor Sir Jack Lewis FRS for his continuing encouragement during the period of writing (much of which was carried out at Robinson College, Cambridge) and also to Drs P. A. Duckworth and P. A. Tasker as well as to several colleagues in the

Department of Chemistry and Biochemistry at James Cook University who provided valuable comments on the text. Finally, I sincerely thank my wife, Fay for her very considerable assistance during preparation of the manuscript.

Townsville, Australia   Leonard F. Lindoy

1

What is different about macrocyclic ligand complexes?

1.1 Background

The understanding of the metal-ion chemistry of macrocyclic ligands has important implications for a range of chemical and biochemical areas. Macrocyclic ligands are polydentate ligands containing their donor atoms either incorporated in or, less commonly, attached to a cyclic backbone. As usually defined, macrocyclic ligands contain at least three donor atoms and the macrocyclic ring should consist of a minimum of nine atoms. The metal-ion chemistry of macrocyclic ligands has now become a major subdivision of inorganic chemistry and undoubtedly great interest in this area will continue in the future.

A very large number of synthetic, as well as many natural, macrocycles have now been studied in considerable depth. A major thrust of many of these studies has been to investigate the unusual properties frequently associated with cyclic ligand complexes. In particular, the investigation of spectral, electrochemical, structural, kinetic, and thermodynamic aspects of macrocyclic complex formation have all received considerable attention.

The fact that macrocyclic ligand complexes are involved in a number of fundamental biological systems has long been recognized. The importance of such complexes, for example to the mechanism of photosynthesis, or to the transport of oxygen in mammalian and other respiratory systems, has provided a motivation for investigation of the metal-ion chemistry of these systems as well as of cyclic ligand systems in general. The possibility of using synthetic macrocycles as models for the biological systems has provided an impetus for much of this research.

There are good reasons for nature choosing macrocyclic derivatives for the important complexes just mentioned – enhanced kinetic and thermodynamic stabilities are bestowed on the respective complexes by the cyclic ligands. The metal ion is thus firmly held in the cavity of the macrocycle such that the biological function of each is not impaired by, for example, competing demetallation reactions.

The porphyrin ring (1) of the iron-containing haem proteins and the related (partially reduced) chlorin (2) complex of magnesium in chlorophyll, together with the corrin ring (3) of vitamin B_{12} have all been studied for many years. However, as well as these conjugated systems, there are a number of other quite different cyclic organic ligands found in nature. An example of this latter group is the antibiotic nonactin (4) which binds potassium selectively and acts as a carrier for this ion across such lipid barriers as cell membranes and artificial lipid bilayers.

Prior to 1960, there existed only one well-established category of synthetic cyclic ligands. These were the highly conjugated phthalocyanines. Phthalocyanine (5) and its derivatives bear a strong structural resemblance to the natural porphyrin systems. The extensive metal-ion chemistry of phthalocyanine ligands is both interesting and varied. For example, specific phthalocyanines have been shown to behave as semiconductors, as catalysts for a variety of chemical transformations, and have been involved in model studies for a number of biochemical systems. Moreover phthalocyanines and related derivatives have been the subject of intense research because of their commercial importance as colouring agents. Thus copper phthalocyanine and its substituted derivatives have found widespread use as both blue and blue-green pigments and dyes (the colour is influenced by the particular substituent present). Apart from their intense colours, the complexes also exhibit marked resistance to degradation: they show high thermal stability, fastness to light, and inertness to acids and alkalis. All these properties favour the use of these compounds as pigments and dyes.

Since 1960, a very large number of other synthetic macrocycles has been prepared and this has resulted in a great increase in interest in all aspects of the chemistry of macrocyclic systems. From about this time there has also been enhanced interest in the role of metal ions in biological systems and many such 'bioinorganic' studies have involved complexes of both natural and synthetic macrocycles. Thus there has been an element of cross-fertilization between these two developing areas, *viz*:

Macrocyclic chemistry

Synthetic macrocycles   ↔   Natural macrocycles

Bioinorganic chemistry

As mentioned already, a considerable amount of the research involving synthetic macrocycles has been directed towards the preparation of

(1)

(2)

(3)

(4)

(5)

model compounds for the natural macrocycles. Although these efforts have not always met with spectacular success, the resultant development of new macrocyclic ligand chemistry has provided a valuable background against which the natural systems can often be seen in clearer perspective.

Apart from the biological implications, aspects of the chemistry of macrocyclic ligands are of relevance to a diverse number of other areas. Indeed, there has been a remarkable expansion of research involving these other areas during recent times. Many of the developments impinge on topics such as metal-ion catalysis, organic synthesis, metal-ion discrimination, and analytical methods, as well as on a number of potential industrial, medical and other applications.

1.2   Steric and electronic considerations
The macrocyclic cavity
Macrocyclic rings and chelate rings. As with simple polydentate ligands, the donor atoms in macrocyclic ligands are normally spaced so that on coordination five-, six-, (and occasionally) seven-membered chelate rings are formed with the metal ion. This requirement results in macrocycles incorporating three donor atoms usually containing between nine and 13 atoms in their inner macrocyclic ring. Thus, ring sizes of between 12 and 17 members most commonly occur when the macrocycle contains four donor atoms, 15–21 members when there are five donor atoms, and 18–25 members for six donor atoms. Examples of different chelate ring patterns for metal complexes of 14-membered macrocycles are given by (6)–(8).

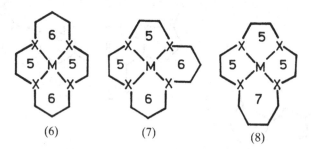

(6)   (7)   (8)

Factors influencing the macrocyclic hole size. The hole size of a macrocyclic ligand is a fundamental structural parameter which will usually influence, to a large degree, the properties of resultant metal complexes relative to those of the corresponding non-cyclic ligands. The large number of X-ray diffraction studies now complete for macrocyclic systems makes it possible to define many of the parameters which affect hole size in such systems. In overall terms, the hole size of a cyclic ligand is very often determined by the number of atoms in the macrocyclic ring. The optimum hole sizes in a series of fully-saturated, tetraaza macrocycles of type (9) in conformations suitable for coordination to a metal ion have

$$
\begin{array}{c}
\text{(CH}_2\text{)}_q \\
\text{H}\overset{\displaystyle}{N} \qquad N\text{H} \\
\text{(CH}_2\text{)}_n \qquad \text{(CH}_2\text{)}_p \\
\text{H}\overset{\displaystyle}{N} \qquad N\text{H} \\
\text{(CH}_2\text{)}_m
\end{array}
$$

(9)

been calculated (Busch, 1978) using the procedures of molecular mechanics. In these calculations the atomic positions corresponding to the minimum strain energies for the respective ligands were derived. The results indicate a regular increment of 0.10–0.15 Å in the mean radius of the central hole for each additional atom in the macrocyclic ring. As calculated, the radii reflect the 'natural' variation of the macrocyclic hole size in the uncomplexed ligands. The procedure does not allow for the ring expansions and contractions which are a feature of coordination to metals having radii which are greater or smaller than the 'natural' hole size of the free ligand.

A simpler procedure for approximating the hole sizes of macrocyclic ligands for comparison purposes has been used to compare the hole size variations which occur for related coordinated macrocycles as the ring size is systematically varied (Henrick, Tasker & Lindoy, 1985). The procedure gives the expected smaller increase in hole size as the number of atoms in the macrocyclic ring is successively increased. Thus, for the complexes of flexible macrocycles, the moderate differences (of about 0.04–0.05 Å) observed for adjacent rings along a given ligand series reflect expansion of the smaller rings and contraction of the larger rings such that the fit for the metal ion is improved in each complex. Nevertheless, it is to be expected that each ring contraction or expansion will be characterized by a concomitant increase in ligand strain energy which may be reflected in less favourable chelate ring conformations or in distortion of the coordination geometry about the metal ion. Thus for such complexes, the observed macrocyclic hole size will reflect a balance between the dictates of the metal ion and those of the macrocyclic ring involved. When the ligand contains a rigid backbone, the capacity for radial expansion or contraction may be severely limited and, under such conditions, metal-donor atom bond distances which are considerably compressed or stretched from their normal values may occur. Thus it has been observed (Hoard, 1975) that the Ni–N bonds in several porphyrin complexes of nickel are longer than occur in related diamagnetic nickel complexes in which a non-constrained N_4-donor set is presented to the nickel.

Apart from the number of atoms in the macrocyclic ring, the nature of the donor atoms may also affect the hole size of a macrocyclic ring. Replacement of a small donor atom by a larger one in a given ligand framework is expected to affect (but not necessarily markedly reduce) the cavity size available to the metal ion. For example, on substitution of sulfur for nitrogen, a partial compensation for the effect of the larger sulfur will occur since the carbon-sulfur bonds in the macrocyclic ring will be longer than the corresponding carbon-nitrogen bonds. Thus, although the bonding cavity of the macrocycle tends to be reduced on substitution of larger donor atoms for smaller ones, this effect may be offset to a lesser or greater degree by the corresponding increase in the 'circumference' of the macrocyclic ring. Such compensation is well-illustrated by comparison of the structure of the complexes of type *trans*-$[NiCl_2L]$ in which L is the N_4-donor macrocycle (10) or the S_2N_2-donor macrocycle (11). Calculations suggest that the bonding cavities in these complexes are very similar in size [and both near ideal for high-spin Ni(II)].

(10)   (11)

The hybridization of the donor atoms may affect the macrocyclic hole size. Relative to the corresponding ligand containing sp^3-amine donor atoms, introduction of sp^2-imine donors into a cyclic ligand normally leads to a reduction of the macrocyclic hole size largely because of decreased 'bites' (distances between adjacent donors) for the chelate rings containing the imine groups. However, because of their higher s-orbital character, the sp^2 hybrids are not as diffuse as sp^3 hybrids and hence their overlap with the appropriate metal orbital leads to a decrease in the corresponding metal-nitrogen bond length. In addition, with sp^2 hybridization there is a prospect of metal to ligand π-bond formation in certain cases – this will further shorten the metal-nitrogen bond. Once again, the two effects are in opposition: although the hole size of the imine-containing macrocycle will be reduced, this will tend to be compensated by the smaller effective covalent radius of the sp^2 nitrogens.

Further considerations. Relative to their open-chain analogues, macrocycles have additional stereochemical constraints resulting from their cyclic nature. These constraints, which depend upon several factors such as the overall macrocyclic ring size and the number and nature of the various chelate rings formed on coordination, will influence the positions of the donor atoms both with respect to each other and to the central metal ion. Such constraints are often also further manifested by a limitation of the possible coordination modes and/or conformations of the coordinated macrocycle. For example, when the metal ion is too large to fit into the available macrocyclic hole, then, provided complexation occurs, the macrocycle will either fold or the metal will be displaced from the donor plane of the ring. Structure (12) is an example of a folded system. For sterically rigid rings, such ligand folding will be energetically unfavourable relative to displacement of the metal ion from the donor plane of the ring. In this case the metal is very often also bound to an axial ligand, as illustrated by (13).

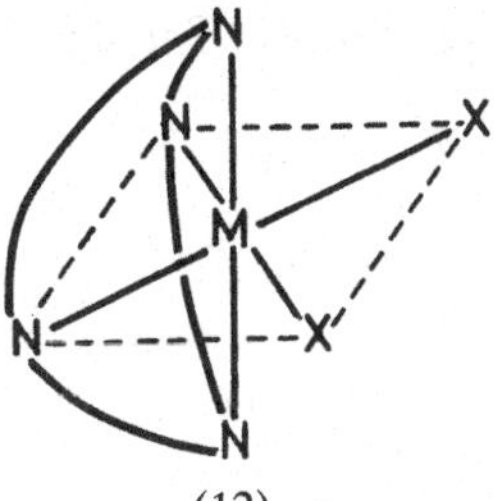

(12)

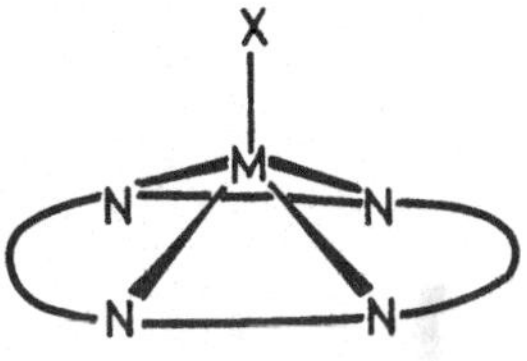

(13)

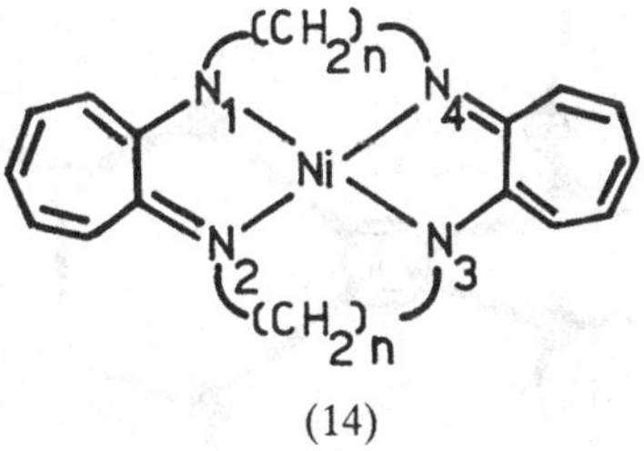

(14)

Macrocycles may also promote the formation of less common coordination geometries for particular metal ions because of increased macrocyclic ring strain on coordination. Such an effect is illustrated by the variation in the structures of the nickel complexes of the 14-, 16-, 18-, and 20-membered 'tropocoronand' macrocycles of type (14) (Imajo, Nakanishi,

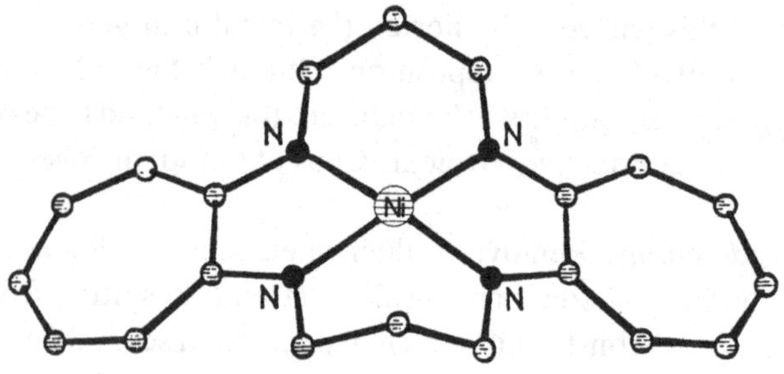

(15)

Roberts & Lippard, 1983). The coordination geometry for the 14-membered ring complex (15) is approximately planar whereas there is a progressive distortion towards a tetrahedral geometry (less common for nickel) as the ring size is increased. The distortion away from planarity is almost certainly a reflection of the increased steric crowding associated with the progressive introduction of larger chelate rings in the series. For the largest (20-membered) ring species (16) an angle of 85° occurs between the N_1–Ni–N_2 and N_3–Ni–N_4 planes.

   Donor atom to metal-ion bond lengths which are shorter or longer than expected as well as unusual angular relationships between such bonds have all been documented in macrocyclic complexes. Such effects can be a prime cause of the unusual properties mentioned previously. Thus the

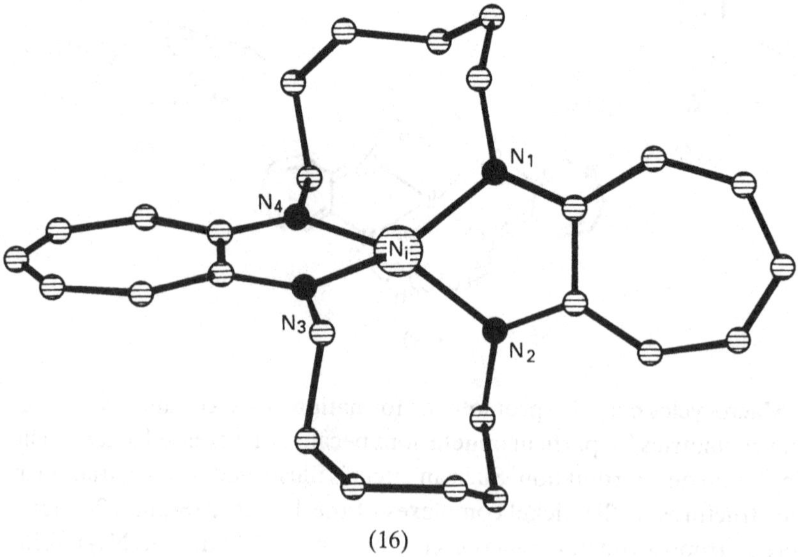

(16)

properties of a particular complex will reflect the compatibility or otherwise of the central cavity for the steric and electronic requirements of the metal ion involved. When a mismatch between the cavity and the metal ion occurs then unusual properties may be generated. The situation in this case approximates that proposed in the 'entatic state' hypothesis (Vallee & Williams, 1968) which relates the enhanced reactivity of metalloenzymes to the unusual coordination geometries of the metal ion observed in many such systems. For macrocyclic complexes, such effects tend to be particularly evident in the results from ligand-field spectrophotometric and electrochemical studies and the resultant theoretical implications have provided the motivation for a considerable number of such investigations in the past. Indeed, in a number of cases the macrocyclic complexes contain the metal in a different electronic ground state compared to the corresponding non-cyclic ligand complexes. Likewise, unusual electrochemical behaviour associated with non-ideal metal-donor bond lengths in cyclic complexes has been well documented. There is, for example, a marked tendency for the smaller macrocyclic rings to stabilize the higher oxidation states of a given metal.

It is important to note that, even when the coordination geometry prescribed by the macrocyclic cavity is ideal for the metal ion involved, unusual kinetic and thermodynamic properties may also be observed (relative to the corresponding open-chain ligand complex). For example, very often the macrocyclic complex will exhibit both enhanced thermodynamic and kinetic stabilities (kinetic stability occurs when there is a reluctance for the ligand to dissociate from its metal ion). These increased stabilities are a manifestation of what has been termed the 'macrocyclic effect' – the multi-faceted origins of which will be discussed in detail in subsequent chapters.

Unsaturation in macrocyclic systems

Consequences of unsaturation. Unsaturation in the macrocyclic ring may have major steric and electronic consequences for the nature of the ring. Extensive unsaturation will result in loss of flexibility with a corresponding restriction of the number of possible modes of coordination. Further, loss of flexibility tends to be reflected in an enhanced 'macrocyclic effect'. For example, if the metal ion is contained in the macrocyclic cavity, the loss of flexibility reduces the possible pathways for ligand dissociation and this tends to increase the kinetic stability of the system. As explained in later chapters, enhanced thermodynamic stabilities will usually also result.

Many macrocycles incorporating high levels of unsaturation have been

(17)   (18)

synthesized and, in extreme cases, the system may be completely conjugated to yield annulene-like rings exhibiting various degrees of complex-
ity such as (17) (Truex & Holm, 1972) and (18) (Ogawa & Shiraishi,
1980). As a consequence of the cyclic character of macrocyclic ligands,
the possibility also exists that a Hückel aromatic system containing
$(4n + 2)\pi$-electrons will occur – the porphyrins form one such group of
ligands. Such aromaticity may thus serve as an additional contribution
towards increasing the difference between a particular cyclic ligand and
its non-cyclic analogue. Apart from the possible effects of increased
rigidity on the complex, the enhanced electron delocalization associated
with rings of this type may also markedly affect the nature of the cyclic
complex formed. The capacity for such rings to act as electron sinks
undoubtedly accounts for the ability of many ligands of this type to
stabilize metal ions in unusual oxidation states. For several systems,
electron spin resonance and other studies have confirmed that substantial
transference of electron density may occur between the metal ion and the
macrocyclic ligand. Although the most favourable overlap of the extensive π-cloud will occur when highly-conjugated macrocyclic ligands are
planar, a number of structural studies have now amply demonstrated
that, for these large ring systems, deviations from planarity can readily
occur. For example, the porphyrins yield metal complexes containing the
ligand in both planar and 'ruffled' non-planar forms. This 'ruffling' of the
porphyrin core has been discussed (Hoard, 1975) in terms of a concomitant shortening of the metal-nitrogen bond lengths in specific complexes.

   Similar distortions from planarity are also observed in a range of
conjugated synthetic ring systems. For example, the tetraaza macrocycle
(19) adopts a saddle-shaped configuration in its complexes with a number
of transition metal ions (Goedken, Pluth, Peng & Bursten, 1976) even
though the four donor atoms remain essentially planar. In this case, the
non-planarity of the remainder of the ligand appears to result largely from
steric clashes between the methyl substituents and the aromatic rings.

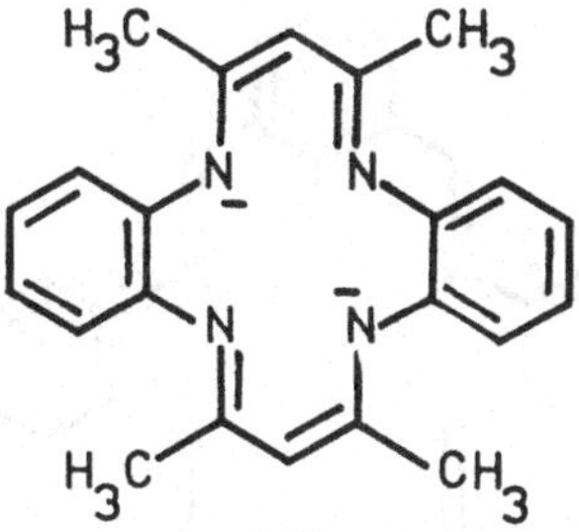

(19)

Unsaturation and less-common coordination geometries. The inherent rigidity of unsaturated macrocycles may also aid the adoption of less-common coordination geometries. For example, the high-spin Fe(II) complex of (20) contains the metal ion in the plane of the macrocycle (which coordinates via five of its seven nitrogens). Overall, the complex has a less-common pentagonal bipyramidal coordination geometry with the apical sites occupied by water molecules (Bishop *et al.*, 1980). Complexes of this and the analogous macrocycle incorporating a bipyridyl moiety (instead of the phenanthroline fragment) have been synthesized for a range of metal ions. The metal coordination geometries are pentagonal bipyramidal or pentagonal pyramidal in most of these complexes.

In a similar manner, the potentially sexadentate macrocycle (21) yields Pb(II) and Cd(II) complexes which have unusual geometries (Drew *et al.*, 1979). With Pb(II), the stereochemistry is hexagonal pyramidal with one axial site occupied by a water molecule and the other filled by a sterically-active lone pair of electrons on the metal ion. The Cd(II) complex is eight-coordinate (with a water molecule and a perchlorate group occupying axial positions); however the cadmium ion is not held centrally in the macrocyclic hole but is displaced to one side, presumably reflecting the

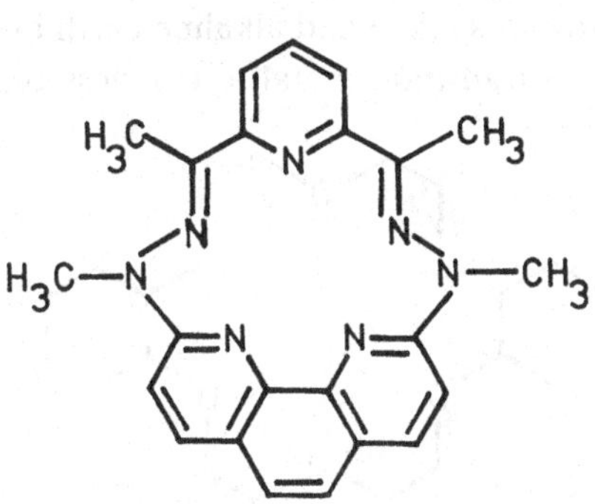

(20)

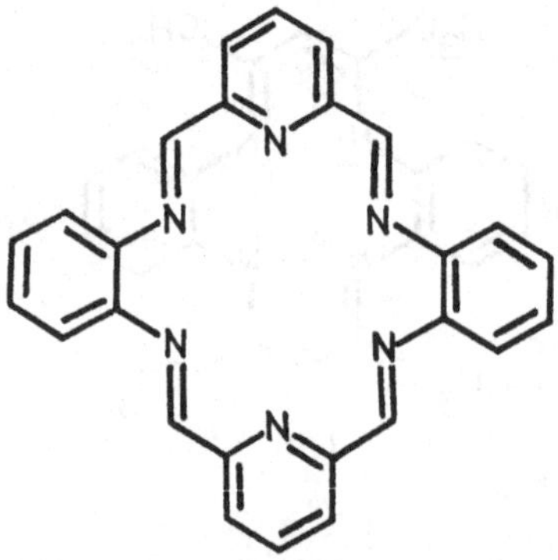

(21)

small size of this ion relative to the available hole size in the macrocycle. In this case the unsaturation in the ligand apparently restricts the 'closing-in' of the donor atoms towards the Cd(II) ion which might be expected to occur for a less rigid cyclic system.

1.3   Some representative macrocyclic systems
Macrocyclic ligands – two subdivisions

Based on donor atom type, macrocyclic ligands can be considered to span two extreme types. First there are those systems which chiefly contain nitrogen, sulfur, phosphorus, and/or arsenic donors. These macrocycles tend to have considerable affinity for transition and other heavy metal ions; they usually show much less tendency to form stable complexes with ions of the alkali and alkaline earth metals. The present discussion will be restricted to a consideration of a selection of such ligands and their complexes.

The second ligand type consists of a large group of cyclic compounds incorporating numbers of ether functions as donors. Structure (22) illustrates a typical example. Such 'crown' polyethers usually show strong complexing ability towards alkali and alkaline earth ions but their tendency to coordinate to transition metal ions is less than for the above

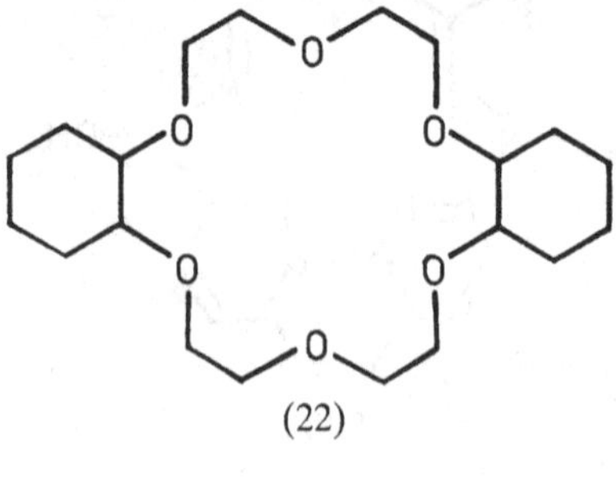

(22)

category. A discussion of the chemistry of the crown polyethers and related ligands is deferred to Chapter 4.

Ring systems incorporating N, S, P or As donors

As background to the material described in the chapters which follow, it is appropriate at this point to introduce a few further examples of particular (monocyclic) ring types. When taken together with the macrocycles described so far, these examples enable some perspective to be gained of the structural diversity for 'simple' rings which now exists. Nevertheless, it is emphasized that collectively these systems represent only a very minor selection from the large number of such rings now reported.

All N-donor systems. The N_3-donor macrocycle (23) has ten atoms in its macrocyclic ring and is too small to completely encircle a metal ion

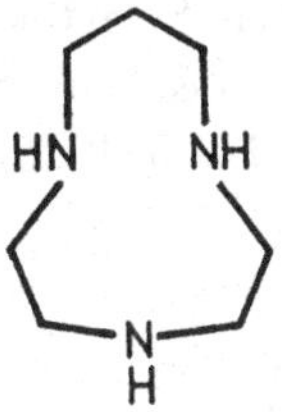

(23)

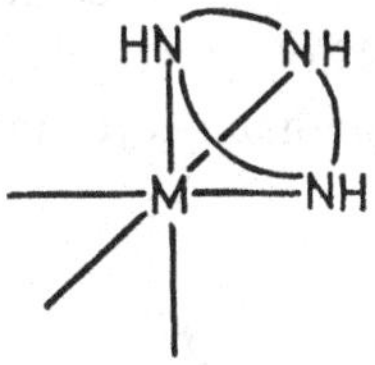

(24)

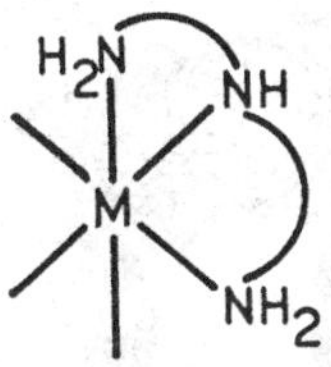

facial
(25)

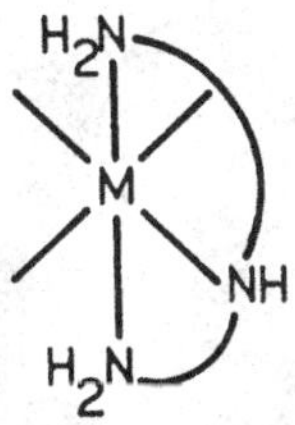

meridional
(26)

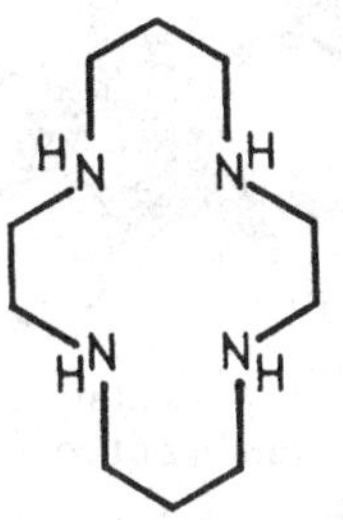

(27)

(Koyama & Yoshino, 1972). This ligand will coordinate such that the metal-ion lies out of the mean plane of the donor atoms and, for example around one face (24) when occupying three coordination positions in an octahedral metal complex. This ligand system is thus restricted to facial coordination in contrast to corresponding open-chain derivatives [which, in principle, may coordinate in either a facial (25) or a meridional (26) manner].

The 14-membered macrocycle 1, 4, 8, 11-tetraazacyclotetradecane (27; cyclam), has been the subject of numerous investigations. Cyclam is large enough to encircle a range of metal ions; for instance, X-ray diffraction studies confirm that *trans*-planar coordination (see Figure 1.1) occurs for this ligand in complexes of Ni(II), Ni(III), Co(II), Cu(II), and Tc(V) (Zuckman *et al.*, 1981). The inherent flexibility of the macrocyclic ring aids its coordination to these metal ions of different radii since ring expansions or contractions may readily occur. Moreover the flexibility also permits this ligand to coordinate in a folded form around, for example, four positions of an octahedral metal ion. Such folded configurations occur in *cis*-[CoCl$_2$(cyclam)]Cl (Robertson & Whimp, 1988) and *cis*-[Co(1,2-diaminoethane)(cyclam)]Cl$_3$ (Lai & Poon, 1976).

Figure 1.1. A space-filling representation of cyclam coordinated in a planar arrangement. In a number of complexes, monodentate ligands also occupy the axial sites.

For coordinated ligands which contain more than one secondary amine, different combinations of the chiral nitrogen donor groups are possible. For cyclam, the different possibilities are illustrated by (28)–(32) (Bosnich, Poon & Tobe, 1965). However, not all combinations will give rise to structures of equal energy since the overall strain energy will be very dependent on the conformations available to the individual five- and six-membered chelate rings in the complex; that is, the conformations which occur for the respective chelate rings are related to the configurations of the nitrogen donors in the structure. For coordination in a plane, the lowest energy structure will result when the five- and six-membered chelate rings have the stable gauche (carbon atoms equally displaced on opposite sides of the corresponding MN_2-plane) and chair (as occurs in cyclohexane) conformations, respectively. Arrangement (30) is the only one which permits both five-membered rings to be gauche and both six-membered rings to be chair – see (33); this structure also corresponds to the space-filling representation given in Figure 1.1. A number of X-ray studies confirm that arrangement (30) occurs experimentally. For the case of a folded ligand complex, the NH group orientations shown in (32) give the most favourable chelate ring conformations.

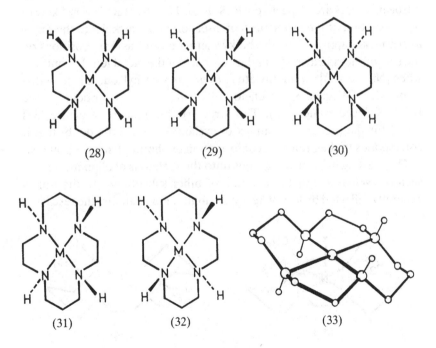

(28)   (29)   (30)

(31)   (32)   (33)

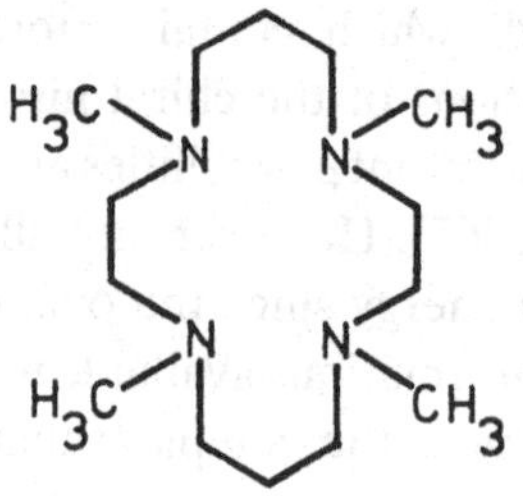

(34)

Unlike cyclam, *N*-tetramethylated cyclam (34) (Barefield & Wagner, 1973) tends to promote the formation of metal complexes which are five-coordinate (Micheloni, Paoletti, Burki & Kaden, 1982). Thus, X-ray diffraction studies of $[NiLN_3]^+$ (D'Aniello *et al.*, 1975) and $[ZnLCl]^+$ (Alcock, Herron & Moore, 1978) where L = (34), both indicate square pyramidal structures for these complexes with the monodentate ligands occupying axial positions; in solution, nmr investigations of the Ni(II) and Zn(II) complexes (Alcock, Herron & Moore, 1978; Herron & Moore, 1979) indicate fluxional behaviour involving likely trigonal bipyramidal species. The nickel complex (which was prepared directly from the free ligand) has structure (35) in which the methyl groups are all directed to one side of the N_4-plane such that the chiral configurations for the nitrogen donors are respectively R, S, R, S. This structure is very likely to be a reflection of the mechanism of insertion of the nickel into the ring – as the nickel approaches, the bulky methyl groups assume positions on the side of the ring away from the approach of the metal ion. In contrast, when $[Ni(cyclam)]^{2+}$ is methylated *in situ* using methyl iodide in dimethyl sulphoxide (Wagner & Barefield, 1976) then the solid product is the R,S,S,R-diastereomer (36). This structure is similar to structure (33) found for the precursor complex of cyclam; there is little doubt that it corresponds to the preferred conformation on thermodynamic grounds.

The grafting of aromatic groups onto the skeletons of aliphatic ligands such as cyclam and/or the addition of other substituents to the ring is generally reflected by loss of ligand flexibility. For example, the restricted

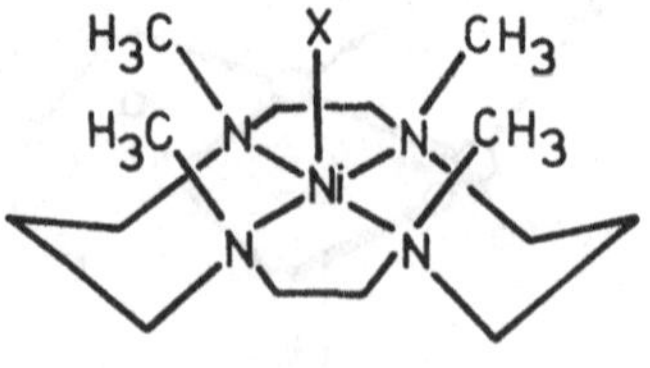

(35)

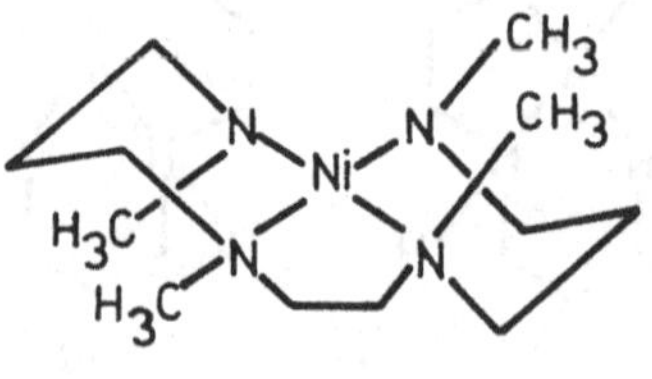

(36)

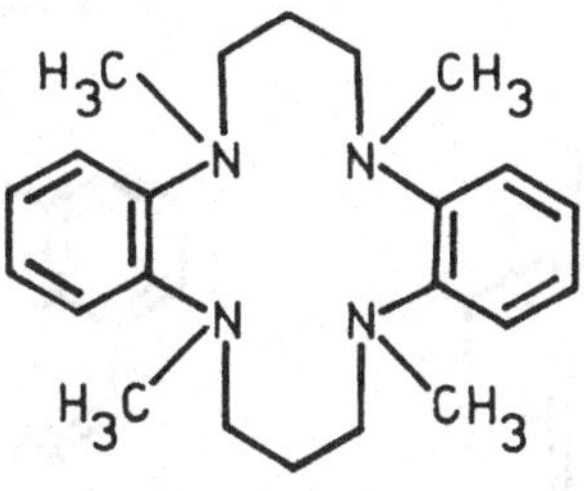

(37)

conformational flexibility of a ligand (37) is manifested in metal-ion complexation behaviour which is slower relative to cyclam (Klaehn, Paulus, Grewe & Elias, 1984).

Emphasis in the early years of macrocyclic chemistry was given to the synthesis of a large number of N_4-donor cyclic systems but synthetic procedures were subsequently developed for obtaining larger-ring species incorporating more than four donors [such as (38) and (39)].

Macrocycle (38) is a member of the family of unsubstituted aliphatic N_n-donor systems [(23) and (27) are others] which have formed the basis of numerous investigations over the years. For the larger rings, such studies became possible following the publication of a general synthesis for this class of macrocycle in 1974 (Richman & Atkins, 1974). Because of their structural simplicity, such rings have been regarded as the parent systems for a number of related (but structurally more complex) macrocycles. In contrast to the flexibility of (38), macrocycle (39) incorporates a 2,2′,2″-terpyridyl fragment in its structure which will contribute to the general rigidity of the ring (Constable *et al.*, 1985). This ligand is potentially quinquedentate since steric restraints will not permit the two nitrogens with methyl groups attached to coordinate to a metal contained

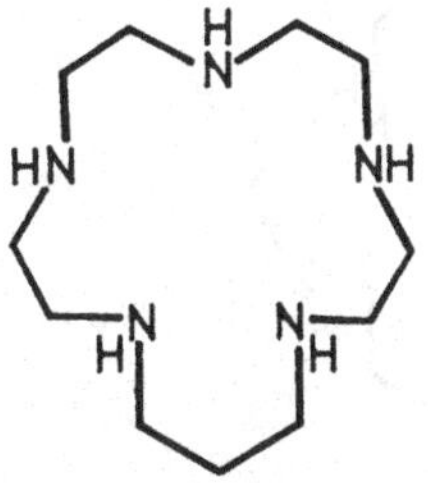

(38)

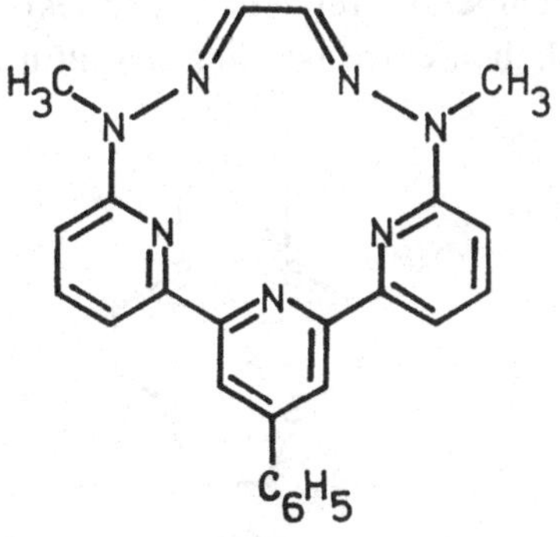

(39)

(40)   (41)

in the ligand cavity. As mentioned previously, ligands of this general type show a tendency to yield complexes which have unusual coordination geometries – a tendency which, in part, reflects the inherent rigidity of such systems. Related all-nitrogen donor macrocycles containing even larger rings have been prepared – these include the Schiff-base derivative (40) (Abid, Fenton, Casellato & Vigato, 1984) and the aesthetically-pleasing hexapyridine derivative (41) (Newkome & Lee, 1983).

Other donor systems. Although not as common as all-nitrogen donor rings, macrocycles incorporating sulfur donor atoms have been widely reported. Structures (42)–(45) illustrate four representative examples of this type.

The S_4 system (42) is the 13-membered ring derivative of a series of related S_4-macrocycles having ring sizes of between 12 and 16 members. As for the corresponding N_4-analogues, this ligand series has been used to investigate the effect of ring size on the coordination behaviour of this ligand type (Pett *et al.*, 1983). Similarly, the mixed donor species (43), is a member of a related N_2S_2-series of macrocycles. These ligands, together with their corresponding *cis*-sulfur isomers, have also been used to study

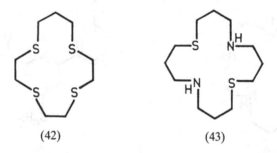

(42)   (43)

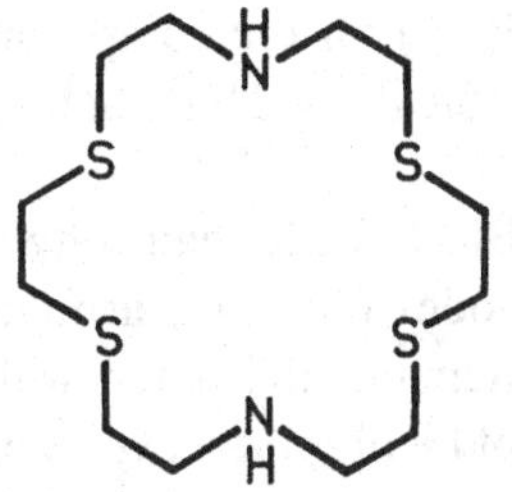

(44)

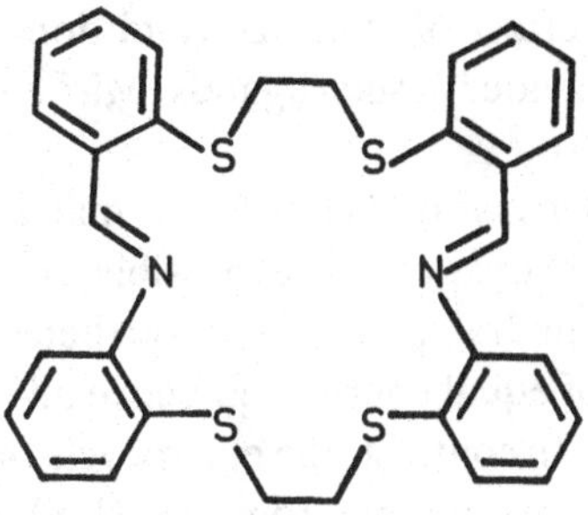

(45)

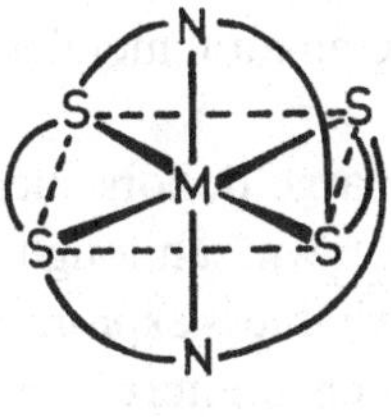

(46)

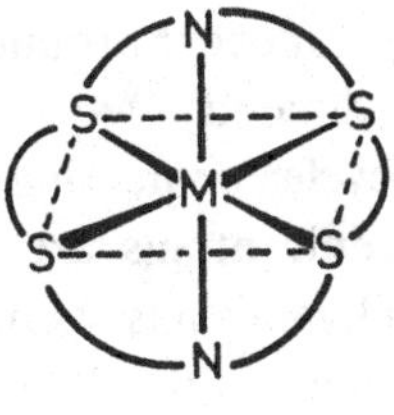

(47)

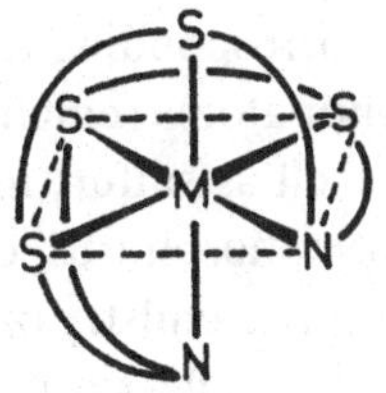

(48)

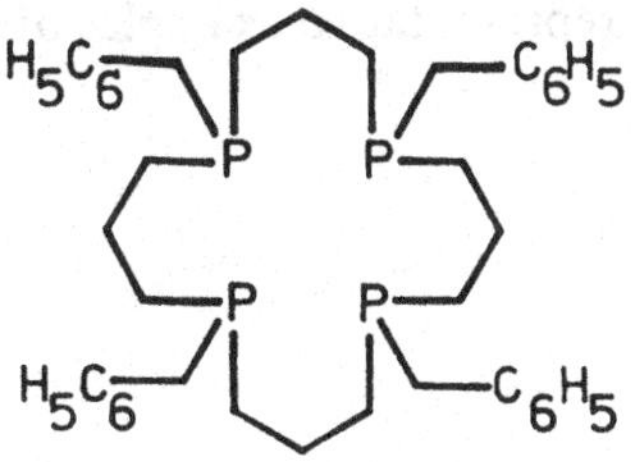

(49)

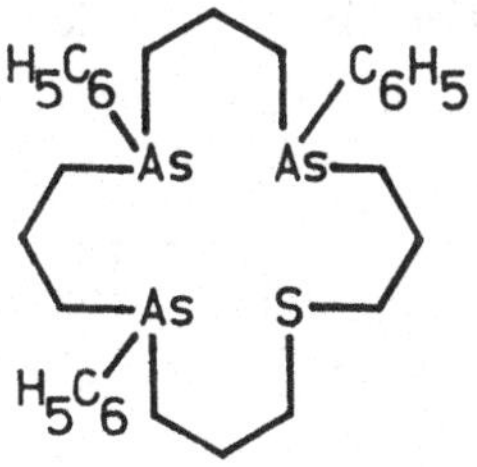

(50)

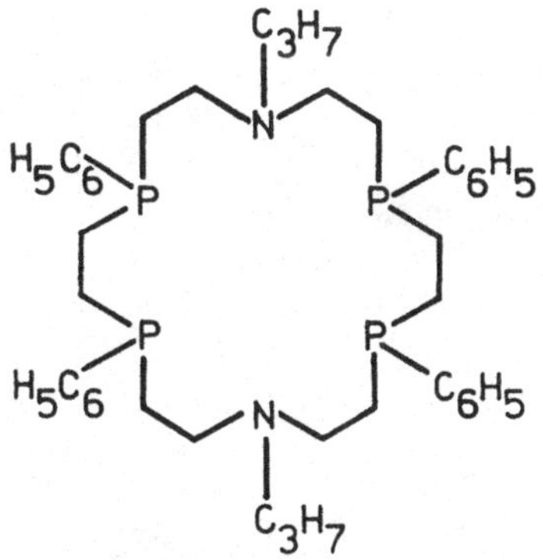

(51)

the effect of ring size (and donor atom position) on the complexation behaviour of such ligands (Micheloni, Paoletti, Siegfried-Hertli & Kaden, 1985).

Because of its flexibility, macrocycle (44) (Black & McLean, 1968) has the choice of three possible coordination modes (46)–(48) around an octahedral metal ion; nevertheless, for a given metal, all isomers will not be of equal energy and hence all three may not be observed experimentally. In contrast, the macrocycle (45) is partially constrained since each set of three donor atoms (S–N–S) is incorporated in a fully conjugated section of the ligand's backbone (Lindoy & Busch, 1969). These two conjugated sections will tend to remain planar such that only a configuration corresponding to (46) can occur around an octahedral metal ion if all donor atoms remain coordinated.

As well as sulfur, macrocycles containing other large donors such as tertiary phosphorus or arsenic atoms are also known, although the metal-ion chemistry of such ligands has been somewhat less explored. In part, this reflects the synthetic difficulties often encountered in the preparation of ligands containing these heteroatoms; structures (49) (Horner, Walach & Kunz, 1978), (50) (Kauffmann & Ennen, 1981), and (51) (Mealli *et al.*, 1985) provide three representative examples of such macrocycles.

2

Synthetic procedures

2.1   Two major synthetic categories

The procedures for synthesizing macrocyclic ligand compounds are many and varied. However, the published cyclizations can be broadly subdivided into two major categories. The first of these are the 'direct' syntheses in which the cyclization proceeds by a conventional organic reaction and does not depend on the directing influence of a metal ion. In the second group of reactions the generation of the cyclic product is influenced by the presence of a metal ion – the metal ion acts as a 'template' for the cyclization reaction. There are a great many template syntheses of macrocyclic products now documented and such *in situ* procedures have proved to be of major synthetic significance over the years. Nevertheless, many macrocyclic systems are formed by multistep procedures and hence may involve a composite of both template and non-template reactions; it may be difficult to define the precise role(s) of the metal ion in such cases. Template effects have been studied in some depth for a few specific systems; however, research directed towards obtaining a more general understanding of such effects has been quite sparse.

A normal priority in both direct and template procedures is to maximize yields of the required product by choosing strategies which inhibit competing linear polymerization and other reactions. Unless special circumstances obtain, polymeric materials are often the major product when macrocyclic syntheses are attempted in the absence of appropriate conditions.

With the aid of a number of examples, some features of both reaction categories are discussed in this chapter. Emphasis in the discussion has been placed on particular examples which illustrate representative cyclizations. It is not possible to discuss all the organic condensation reactions which may, in theory, be employed to produce cyclic ligands. Indeed, such a discussion would need to encompass a large proportion of the entire range of synthetic organic reactions now reported.

2.2   Direct macrocycle syntheses
High-dilution procedures

A typical direct synthetic procedure involves the reaction, in equimolar concentrations, of two reagents incorporating the required fragments for the target macrocycle such that a 1:1 condensation occurs. Such reactions are frequently performed under high-dilution conditions which tend to favour cyclization by enhancing the prospect of the 'half-condensed' moiety reacting with itself 'head-to-tail' rather than undergoing an intermolecular condensation with another molecule in the reaction solution. If the latter occurs then this is the initial step of an oligomerization or polymerization process which will not produce the required monocyclic product.

Many preparations of the polyether class of macrocycles depend on high-dilution procedures for their success (some typical syntheses are given in Chapter 4). An apparatus for accurately controlling the experimental conditions during such syntheses has been developed (Karbach, Löhr & Vögtle, 1981). In a typical preparation, two motorized burettes are arranged to dispense metered quantities of reactants into the reaction solvent at a very slow rate. As a consequence, the concentrations of unreacted reagents in the solution at any given time is extremely small. Reactions of this type may proceed over several days but, for favourable systems, spectacularly high yields may result.

Procedures involving moderate to low dilution

A range of direct syntheses have been performed under conditions of moderate to low dilution but have still led to isolation of the required cyclic product in reasonable to high yields. In some of these so-called direct syntheses, ions such as sodium or potassium were present in the reaction solution. It has often been suspected (and has been documented for a number of systems) that these ions may also undertake a template role. In this manner, such syntheses may fall between the extreme direct and *in situ* categories mentioned at the beginning of this chapter. There is little doubt that such an effect is the source of the larger-than-expected yields obtained in some syntheses of this type. Nevertheless, this is not always the case and there are further mechanisms for assisting cyclization which may operate for particular systems. For such systems, special circumstances usually appear to be present. Examples of this latter type are discussed later in this chapter.

Some representative syntheses

Cyclizations best performed at high dilution. Equation [2.1] outlines the preparation of a 14-membered S_4-donor macrocycle (52) using a direct procedure. The yield of (52) is quite dependent on the degree of dilution under which the condensation is performed (Rosen & Busch, 1969; Travis & Busch, 1974): initially a yield of $7\frac{1}{2}\%$ was obtained but this was subsequently increased to 55% by performing the reaction at higher dilution. The 2:2 condensation product containing eight sulfur atoms in the ring was shown to be one of the side products formed under the conditions employed (Travis & Busch, 1970). This example serves to illustrate some of the difficulties which may be associated with the use of direct condensation reactions for macrocycle synthesis. In general, such difficulties may include some or all of the following: the formation of undesirable side products, the need to use large volumes of solvents, and variable yields (arising from lack of stereochemical control during the cyclization step).

[2.1]

(52)

A number of phosphorus- and arsenic-containing macrocycles have been obtained by direct means. For example, using high-dilution procedures (with tetrahydrofuran as solvent) and starting from the appropriate precursor (53), 14-membered rings of type (54) containing either four tertiary arsenic or two arsenic and two phosphorus donors were obtained [2.2]. These macrocycles are able to exist in a number of isomeric forms since the configurations about the respective heteroatoms are not readily interconvertible. For both systems, quite poor yields were reported (Kyba & Chou, 1981). Other preparations using the appropriate

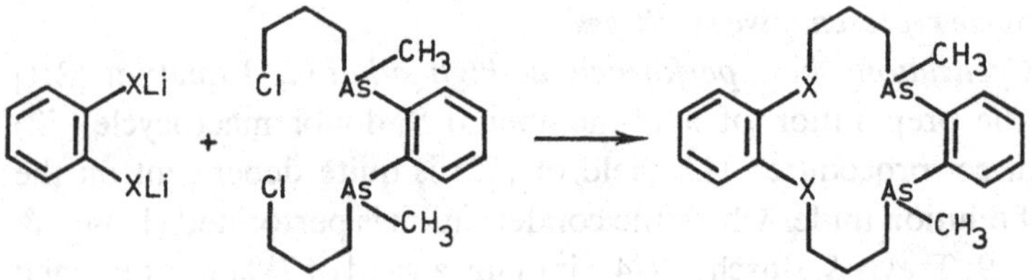

$$[2.2]$$

(53)

X = AsCH₃ or PC₆H₅

(54)

precursors also led to products of type (54) but with X = O or S. The yields for these latter syntheses were 34% (X = O) and 37% (X = S).

The above syntheses constitute an extension of earlier work in which precursors of type (53, X = AsCH₃ and PC₆H₅) were employed for the synthesis of a range of three-donor cyclic systems of general structure (55) (Kyba & Chou, 1980). Yields for these reactions ranged from 19 to 40%. The arsenic derivatives of type (55) were the first macrocycles incorporating this heteroatom to be reported.

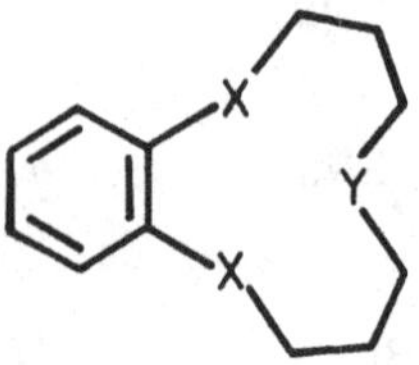

X = AsCH₃; Y = AsC₆H₅, PC₆H₅, S,
                    NCH₃, O

X = PC₆H₅; Y = AsC₆H₅

(55)

Cyclizations at moderate to low dilution. A series of *N*-tosylated (tosyl = *p*-toluenesulphonyl; Ts) macrocycles may be readily prepared by direct means starting from pre-tosylated reactants (Richman & Atkins, 1974). Equation [2.3] summarizes an example of this useful reaction type. Reasonable yields (often considerably better than 50%) of such cyclic tosylated products may be achieved in spite of the fact that such reactions are usually performed at moderate dilution. A number of procedures exist for detosylation of these products to yield the corresponding rings containing only secondary nitrogens; a common method has been to treat the tosylated intermediate with hot concentrated sulfuric acid for several days.

Despite the non-use of high-dilution conditions, the reasonable yields of cyclic products from syntheses of the type illustrated by [2.3] are believed to stem, in part, from the influence of the bulky tosyl groups.

$$[2.3]$$

These groups will reduce the number of conformational degrees of freedom (such as bond rotation) in the reactants and/or intermediates. It is this reduction which is thought to facilitate cyclization relative to polymerization for these systems; in essence, the enhancement of cyclization can be considered to be largely a consequence of favourable entropy effects.

Other types of macrocyclic rings sometimes also form in high yield without recourse to 'metal template' or 'high dilution' techniques. For such systems, special circumstances usually appear to be present. For example, under conditions of moderate dilution, the N_4-macrocycles of type (56), containing ring sizes of between 14 and 22 members are readily obtainable by direct condensation – see [2.4] (Owston *et al.*, 1980). Yields of between 73% and 94% were reported for these products. In this ligand series the presence of anilino-nitrogen atoms allows intramolecular hydrogen bonding between pairs of nitrogen donors: such hydrogen bond formation will reduce considerably the lone-pair repulsions which would otherwise occur within the macrocyclic ring. Thus, in this system, the hydrogen bonds can be thought of as acting like a metal cation in both aiding the formation of the ring and stabilizing it once it is formed.

$$[2.4]$$

(56)

A similar consideration is very likely also to apply to the reaction illustrated by [2.5] involving the sterically rigid precursors, 2,9-dichloro-1,10-phenanthroline and 2,9-diamino-1,10-phenanthroline (Ogawa, Yamaguchi & Gotoh, 1974). The product macrocycle (57) was obtained

$$[2.5]$$

(57)

by heating these precursors in nitrobenzene in the presence of potassium carbonate which acts as an acid scavenger. The product was isolated as yellow needles in greater than 90% yield even though the condensation was performed in the absence of high-dilution conditions. Once again, the high yield probably reflects the formation of a hydrogen-bonded intermediate in which intramolecular NH···H bridges occur between heterocyclic nitrogen atoms such that the overall configuration favours ring closure. Similarly, the inherent rigidity of both reactants may also result in an entropic contribution which favours cyclization. Indeed, an alternative preparation of (57) provides the ultimate example of non-high-dilution cyclization (Ogawa, 1977). If 2,9-diamino-1,10-phenanthroline is heated slightly above its melting point in the absence of solvent, then ammonia is released and (57) is formed in high yield!

Further comments

The preceding discussion presents a modest selection from the large number of different types of direct syntheses found in the literature. The examples were selected to illustrate some of the factors which may effect a particular cyclization reaction in the absence of a template. Because of the large diversity of reaction types and conditions reported for such reactions, little purpose is served by presenting further examples here. As mentioned already, to do so would largely result in a catalogue of organic reaction types, usually performed under one or other of the reaction conditions just discussed. On the other hand, the use of a metal template is a procedure largely confined to the synthesis of multidentate ligands and especially of macrocyclic ones. As such, it is appropriate that template reactions be given greater prominence in any discussion of macrocyclic ligand syntheses. Accordingly, a moderately comprehensive selection of template reaction types is presented in the following section.

2.3   Metal-ion template syntheses
Some early syntheses

The effect of metal ions in promoting certain cyclization reactions has been recognized for a very long time. Thus the first synthetic

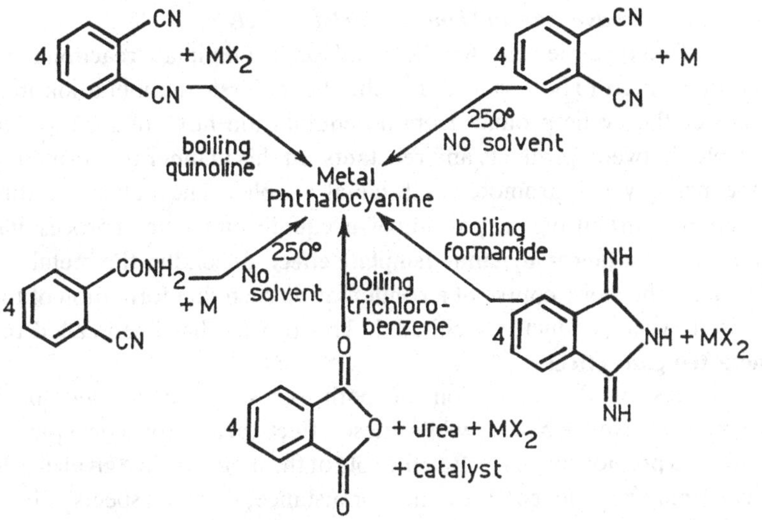

Figure 2.1. Typical *in situ* syntheses for metal phthalocyanines.

macrocycle was obtained by a template reaction in 1928 as a side product formed during the preparation of phthalimide by reaction of phthalic anhydride and ammonia in an iron vessel. This dark blue compound was later shown to be an Fe(II) complex of the highly conjugated macrocycle phthalocyanine (5) which, as mentioned in Chapter 1, bears a strong structural resemblance to the natural porphyrin systems. Since this initial discovery, a variety of *in situ* methods have been employed to prepare phthalocyanine complexes (Moser & Thomas, 1963; Lever, 1965) although there is no single method which can be used for all complexes. Common procedures have involved the reaction between phthalonitrile (or a derivative) and a finely divided metal, metal hydride, oxide or chloride in either the absence or the presence of solvent (see Figure 2.1). Under conditions such as these, the required complexes are very often formed in high yield reflecting the ease with which the cyclization reaction proceeds; however, details of the template role of the metal of such syntheses remain little understood.

Apart from this one-reaction type, the routine use of metal template procedures for obtaining a wide range of macrocyclic systems stems from 1960 when Curtis discovered a template reaction for obtaining an isomeric pair of Ni(II) macrocyclic complexes (Curtis, 1960). Details of this reaction are discussed later in this chapter. The template synthesis of these complexes marked the beginning of renewed interest in macrocyclic ligand chemistry which continues to the present day.

Thermodynamic and kinetic template effects

Two possible roles for the metal ion in a template reaction have been delineated (Thompson & Busch, 1964). First, the metal ion may sequester the cyclic product from an equilibrium mixture such as, for example, between products and reactants. In this manner the formation of the macrocycle is promoted as its metal complex. The metal ion is thus instrumental in shifting the position of an equilibrium – such a process has been termed a **thermodynamic template effect**. Secondly, the metal ion may direct the steric course of a condensation such that formation of the required cyclic product is facilitated. This process has been called the **kinetic template effect**.

Examples of the operation of both types of effect have been documented. Nevertheless, while these effects are useful concepts, as mentioned previously, very often the role of the metal ion in a given *in situ* reaction may be quite complex and, for instance, involve aspects of both effects. As well, the metal may play less obvious roles in such processes. For example, it may mask or activate individual functional groups or influence the reaction in other ways not directly related to the more readily defined steric influences inherent in both template effects.

In view of the above, it is not surprising that a detailed understanding of the part played by the metal ion in many (perhaps the majority!) of published template reactions still remains to be elucidated. Nevertheless, in the following discussion an attempt has been made to illustrate a representative cross section of such *in situ* reactions even though, for many examples, little comment can be made concerning mechanistic details of the respective condensations.

The Curtis synthesis

As mentioned already, Curtis reported the first (Curtis, 1960) of a number of pioneering template reactions for macrocyclic systems which were published in the period 1960 to 1965. In the Curtis synthesis, a yellow crystalline product was observed to result from the reaction of $[Ni(1,2\text{-diaminoethane})_3]^{2+}$ and dry acetone. This product was initially thought to be a bis-ligand complex of the diimine species (58). However, the stability of the product in the presence of boiling acid or alkali was

(58)

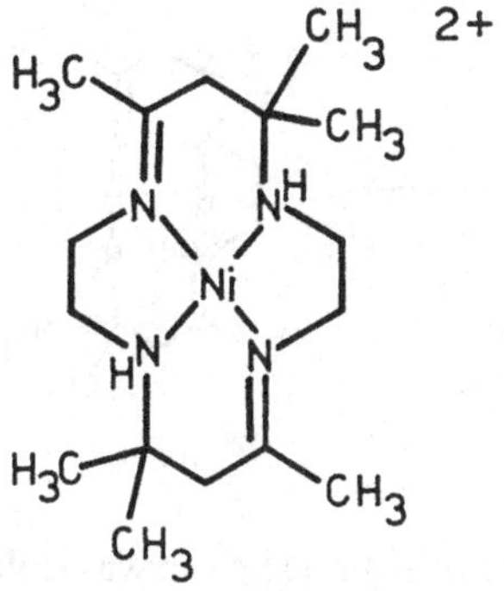

(59) 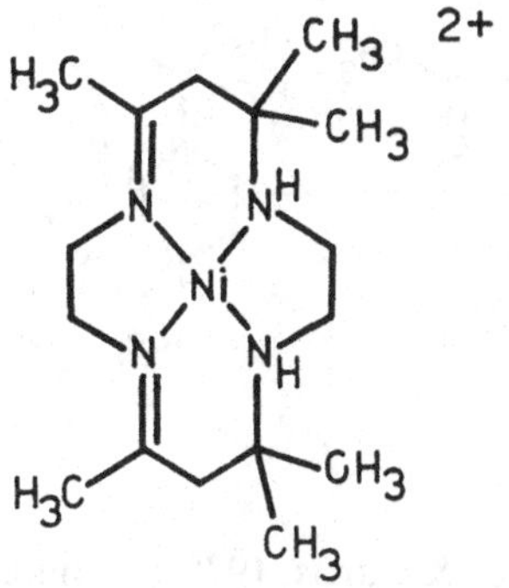   (60)

inconsistent with it containing two ligands of this type. Metal complexes of (58) would be expected to be decomposed under much milder conditions than these. The yellow product was subsequently shown (Curtis & House, 1961; Curtis, Curtis & Powell, 1966) to be a mixture of the isomeric macrocyclic complexes (59) and (60). In this remarkable cyclization reaction, formation of the bridges between the two 1,2-diaminoethane moieties involves condensation of two acetone molecules per bridge.

Although the sequence of the reaction steps remains uncertain, the mechanism may involve the nucleophilic attack of an acetonyl carbanion on the carbon of a coordinated imine as in [2.6].

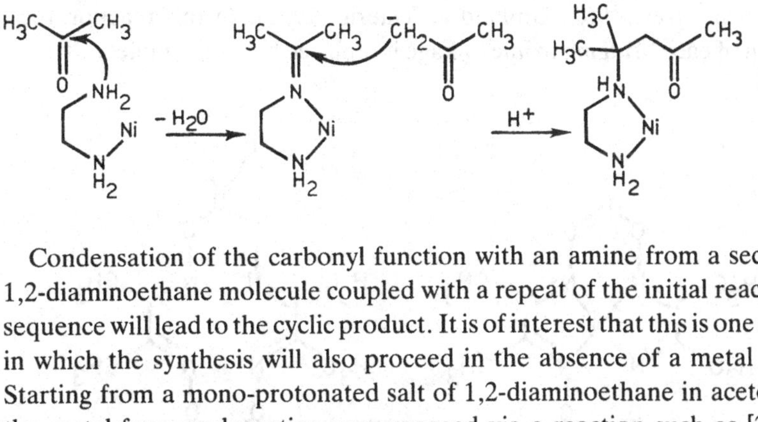

[2.6]

Condensation of the carbonyl function with an amine from a second 1,2-diaminoethane molecule coupled with a repeat of the initial reaction sequence will lead to the cyclic product. It is of interest that this is one case in which the synthesis will also proceed in the absence of a metal ion. Starting from a mono-protonated salt of 1,2-diaminoethane in acetone, the metal-free condensation may proceed via a reaction such as [2.7]. Once again, a hydrogen-bonding network may act as a template for the reaction and also serve to stabilize the product once it is formed. A revised synthesis of the metal-free ligand (61) has been published (Hay,

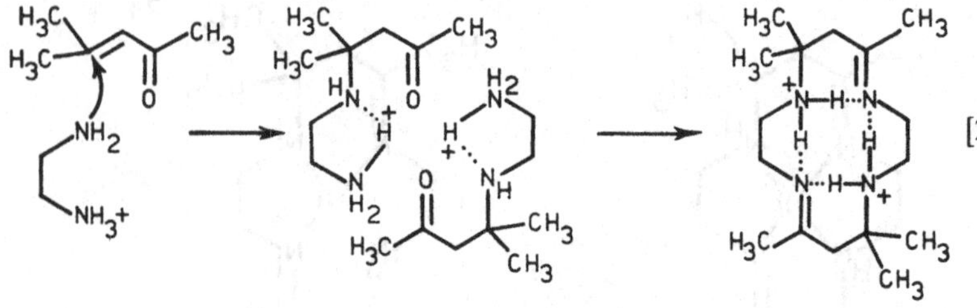

[2.7]

(61)

Lawrance & Curtis, 1975); in this case the *trans*-diene ligand was isolated, for example, in greater than 80% yield, as its dihydrobromide.

A selection from the large number of template reactions published following the original report by Curtis will now be described. Schiff-base and related condensations have figured prominently in these reactions. For ease of presentation, it is convenient to separate examples involving non-Schiff-base condensation from those involving Schiff-base formation. The next two subsections are devoted to descriptions of examples from each of these respective types.

Non-Schiff-base template reactions

Examples of template procedures in which the ring-closing condensation involves reaction at a centre other than a donor atom have been documented. In an early synthesis of this type, the reaction of bis-(dimethylglyoximato)nickel(II) (62) with boron trifluoride was demonstrated to yield the corresponding complex of the N_4-donor macrocycle (63) (Schrauzer, 1962; Umland & Thierig, 1962). In this reaction the proton of each bridging oxime linkage is replaced by a BF_2^+ moiety. X-ray

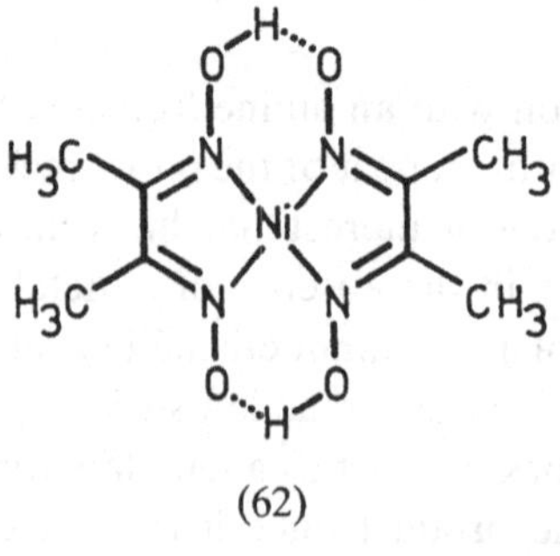

(62)

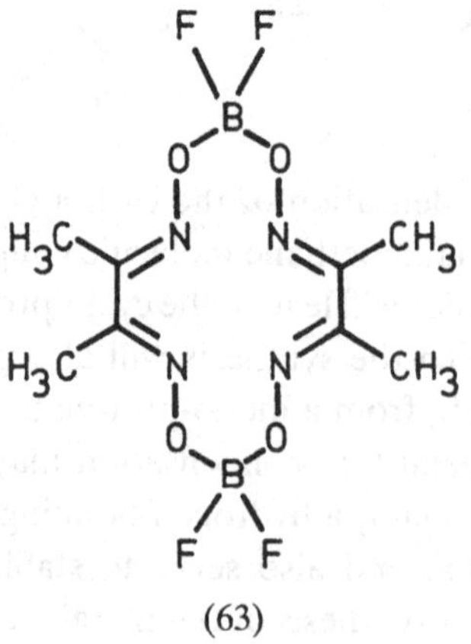

(63)

$$[2.8]$$

studies (Stephens & Vagg, 1977) subsequently confirmed that the complex had the planar structure originally proposed.

A different example of this general type involves the *cis*-dihydrazone complex (64). This complex is prepared by condensation of the corresponding dialdehyde with hydrazine in the presence of nickel or copper ion (Curtis, Einstein & Willis, 1984). The two uncoordinated $-NH_2$ groups react with a ketone or a diketone such that cyclization occurs in each case [2.8]. In these systems, the pendant $-NH_2$ groups are expected to be reactive since their non-coordination results in the respective nitrogen lone pairs being readily available for nucleophilic reaction.

A cyclization reaction involving a 'half-formed' bridge in which alkyl halide functions interact with (initially) coordinated oxygen atoms is illustrated by [2.9] (Kluiber & Sasso, 1970). The X-ray structure of the red paramagnetic nickel complex (65) indicates that the macrocycle coordi-

$$[2.9]$$

(65)

nates in a planar fashion around the Ni(II) with the axial coordination sites occupied by iodide ligands (Johnston & Horrocks, 1971).

Reaction of (66) with the difunctional alkylating agent, α,α-dibromo-*o*-xylene, results in ring closure to produce (67) by bridging the *cis*-thiolato functions of (66) as shown in [2.10] (Thompson & Busch, 1964). This

(66)   (67)   [2.10]

reaction has been proposed to proceed via a kinetic template mechanism in which, after the initial alkylation of one of the coordinated sulfur donors, there is very fast ring-closing condensation involving alkylation of the second sulfur. Kinetic measurements are in accordance with this proposal (Blinn & Busch, 1968): the fast ring closure is a reflection of the necessary proximity of the reacting bromo functional group and the remaining thiolato group once the initial alkylation step is complete. Thus the metal ion directs the steric course of the reaction. It should be noted that if the reaction proceeds as discussed then it is necessary that the coordinated sulfur atoms retain some of the nucleophilic character of the free thiolato ion (even though charge neutralization on coordination to a positive metal ion would be expected to reduce their nucleophilicity). Thus the prospect that the respective condensations may involve sulfur functions which, to a greater or lesser degree, have moved away from the influence of the nickel during the course of the reaction cannot be ruled out.

An X-ray structure (Fernando & Wheatley, 1965) of (66) indicates that the presence of three fused five-membered chelate rings has resulted in the sulfur atoms being somewhat further apart than would occur if, for example, one six-membered chelate ring was present in the structure. As a consequence, the use of a ring-closing reagent containing a four-carbon bridge appears near-ideal for this reaction. Subsequently, this reaction type was utilized successfully to prepare an early example of a phosphorus-containing macrocycle – see [2.11] (Marty & Schwarzenbach, 1970).

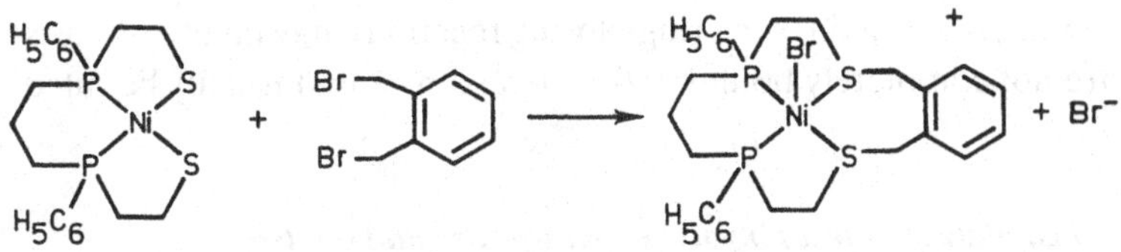

[2.11]

Bridge formation between *cis* functional groups of the type just discussed is not restricted to *cis*-thiolato compounds. Thus, ring-closing
reactions involving *cis* nitrogen functions and *cis* phosphino functions
have been documented; examples are illustrated by [2.12] (Uhlemann &
Phath, 1969), [2.13] (Ansell *et al.*, 1985) and [2.14] (DelDonno & Rosen,
1978). As in these reactions, alkylation of coordinated donor atoms will
only be achieved if these atoms have available a non-bonding pair of
electrons which are suitably orientated to undergo reaction with an
approaching electrophile. However, even if this condition is met, the

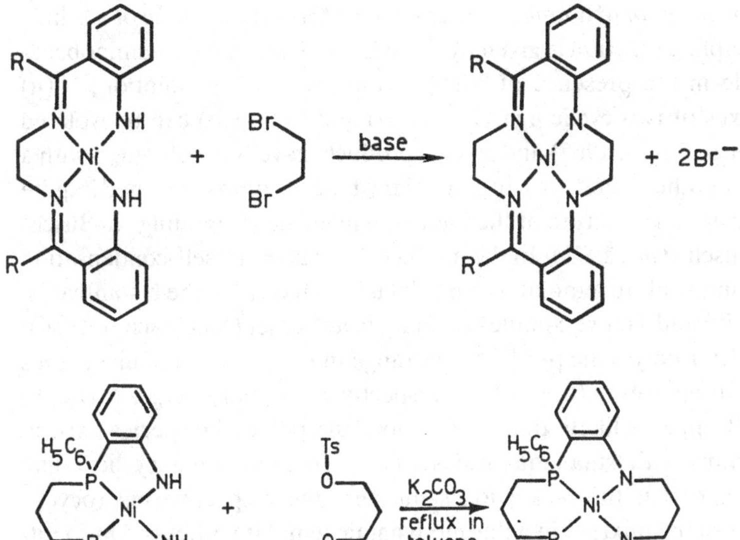

[2.12]

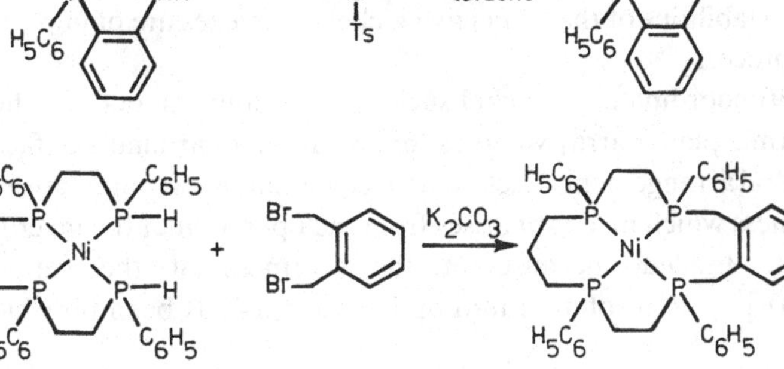

[2.13]

[2.14]

prospect that, once again, such ring-closing reactions may involve donors which are not completely bound to the metal ion cannot usually be ruled out.

Template syntheses involving Schiff-base and related condensations

As mentioned previously, Schiff-base condensations between amines and aldehydes or ketones have played a prominent role in metal-ion template chemistry. A generalized Schiff-base condensation (to produce an imine linkage) is given by [2.15].

$$R^I-H_2N: \overset{R}{\underset{\overset{\|}{O}}{\underset{M^{n+}}{\overset{C\sigma+}{\diagup}}}} \quad \rightleftharpoons \quad \underset{R^I}{\overset{HO-C}{\underset{H-N}{\diagdown}}}\underset{M^{n+}}{\overset{R}{\diagup}} \quad \overset{-H_2O}{\rightleftharpoons} \quad \underset{R^I}{\overset{C}{\underset{N}{\overset{\|}{\diagdown}}}}\underset{M^{n+}}{\overset{R}{\diagup}} \qquad [2.15]$$

Self-condensation of o-aminobenzaldehyde. An early example of a Schiff-base template reaction is given by the self-condensation of *o*-aminobenz-aldehyde in the presence of Ni(II). From the reaction solution, Ni(II) complexes of two cyclic ligands (68; tri) and (69; taab) can be isolated (Melson & Busch, 1965) and similar products have been obtained with a number of other metal ions; in general the type of complex formed is quite dependent on the nature of the metal ion involved (Cummings & Busch, 1970; Busch *et al.*, 1971). In the absence of metal ions, self-condensation is slow and leads to a mixture of products which include the bis-anhydro-trimer (70) and a more complex trisanhydrotetramer (McGeachin, 1966). These latter polycyclic products rearrange in the presence of nickel ions to yield complexes of (68) and (69) respectively (Taylor, Vergez & Busch, 1966). It appears likely that, in solution, the polycyclic species exist in equilibrium with small quantities of the corresponding cyclic imine ligands and that the nickel ion sequesters the respective macrocyclic forms from the mixture in a thermodynamic template process. Undoubt-edly the high stabilities of the respective nickel complexes are of import-ance to this process.

Ligand (69) coordinates to nickel such that the four donors and the metal ion form a planar array whereas (68) coordinates around one face of an octahedral arrangement. Each complex type exhibits a characteristic kinetic inertness which no doubt arises from the operation of the macro-cyclic effect. Indeed, because of the inertness of the cation $[Ni(tri)(H_2O)_3]^{2+}$, its resolution into optical isomers has been possible

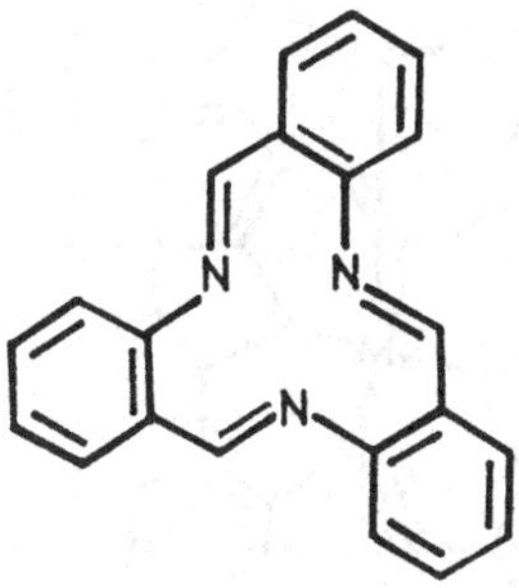

(68)

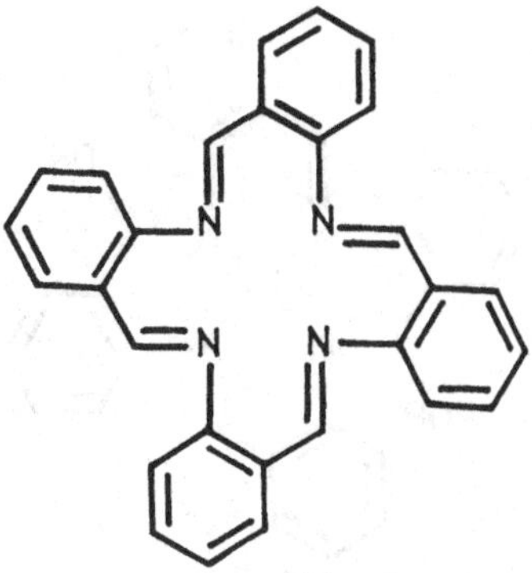

(69)

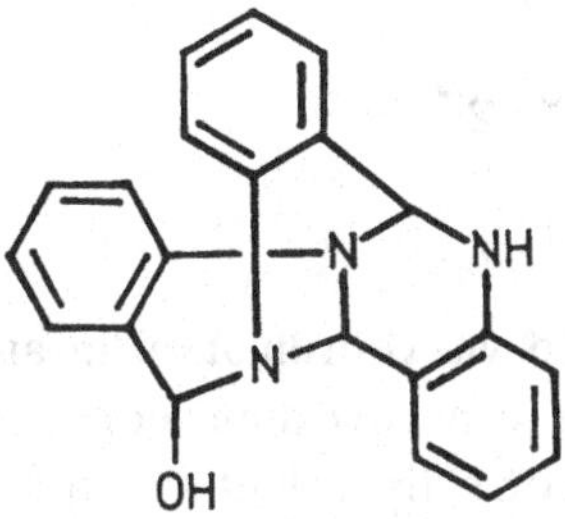

(70)

(Taylor & Busch, 1967). In solution, these isomers are quite stable towards racemization – a property not normally characteristic of high-spin nickel complexes containing open-chain ligands.

It is of interest that (69) does not lie completely flat on coordination to Ni(II) but assumes a conformation in which the aromatic rings point alternately above and below the N_4-plane. The ruffling seems to be a reflection of strain in the respective chelate rings and may be associated with this 16-membered ring being slightly too large for Ni(II) when coordinated in a planar fashion.

The coordinated macrocycle readily reacts with alkoxide ions to yield products of type (71) (Taylor, Urbach & Busch, 1969). In so doing additional flexibility is imparted to the ring which may reduce ring strain and, in part, provide a driving force for the reaction. Thus the coordinated imine carbons appear predisposed to attack by such nucleophiles. Based on this knowledge, elegant template syntheses of three-dimensional derivatives have been performed. The syntheses involved the reaction of [M(taab)]$^{2+}$ (M = Ni, Cu) with the dialkoxide ions derived from bis(2-hydroxyethyl)sulphide or bis(2-hydroxyethyl)methylamine (Katovic, Taylor & Busch, 1969). The products were demonstrated to be monomeric square-pyramidal complexes of type (72). The condensation

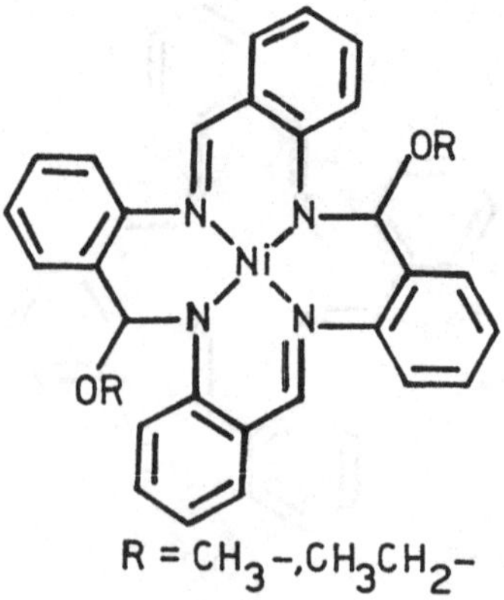

R = CH₃–, CH₃CH₂–

(71)

M = Ni, Cu
X = S, NCH₃

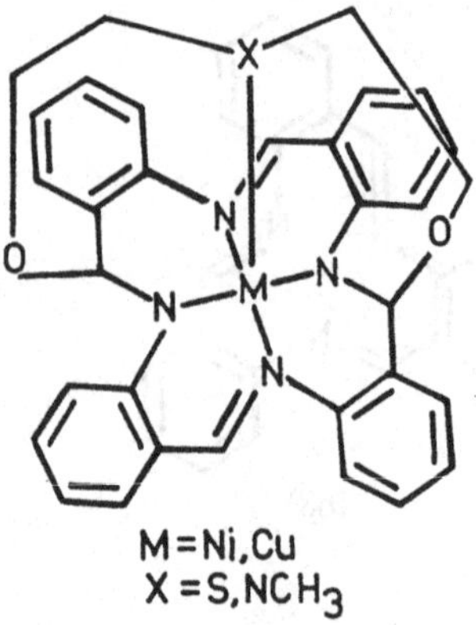

(72)

was proposed to proceed via coordination in an axial position of the centre donor atoms of the respective dinucleophiles such that the terminal alkoxide groups were correctly orientated for attack at *trans* imine carbons in the coordinated macrocyclic ring. In this manner, possible linear polymerization reactions were avoided.

Other template cyclizations. In another Schiff-base template reaction, 1,3-diaminopropane monohydrochloride was reacted with biacetyl in methanol in the presence of Ni(II) to yield the nickel complex of the corresponding cyclic tetraimine – see [2.16] (Jackels *et al.*, 1972). The success of the procedure illustrated is quite dependent on the reaction conditions employed. Attempts to isolate the metal-free macrocycle were unsuccessful – this once again emphasizes the stabilizing role of the metal

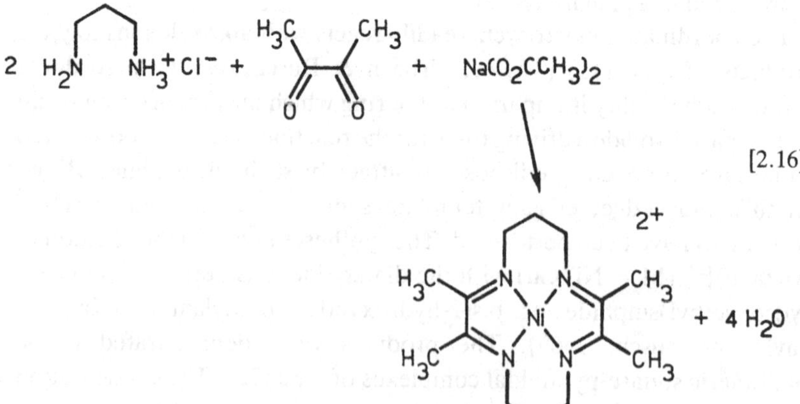

[2.16]

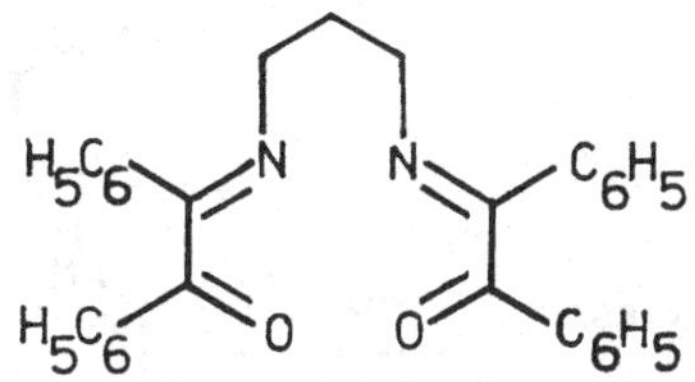

(73)

in such systems. Indeed, it may be relevant that two molar equivalents of benzil will react with 1,3-diaminopropane (slightly more than one molar equivalent) to yield (73). This species only reacts with a further molar equivalent of diamine when a metal such as Co(II) is present; thus, at least for this system, a metal ion is a requirement for ring closure to occur (Welsh, Reynolds & Henry, 1977).

Beginning in 1964, the first of a series of studies of the reaction between β-ketoiminato complexes of type (74) and diamines (o-phenylenediamine, 1,2-diaminoethane, or 1,3-diaminopropane) was published (Jäger, 1964). The products obtained were macrocyclic complexes of type (75). Further members of the series were published subsequently (Jäger, 1969). Typically, Jäger used forcing conditions to effect ring closure such as heating the appropriate precursor complex of type (74) with the required diamine in the absence of solvent. Emphasis in these studies has been given to the synthesis of square-planar Ni(II) and Cu(II) complexes. The cyclization reaction was found to be quite sensitive to the nature of the substituents present in the β-positions of (74). Treatment of the Cu(II) complexes of some of these cyclic ligands with H_2S results in removal of the copper as CuS and the metal-free macrocycle may be isolated. More

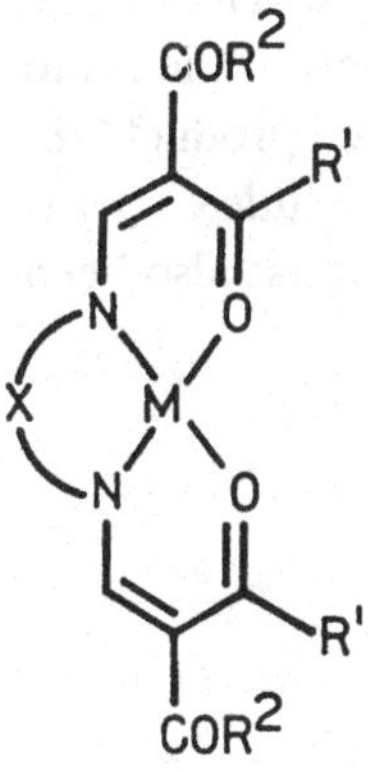

$$X = -(CH_2)_2-, -(CH_2)_3-, o-C_6H_4$$
$$Y = -(CH_2)_2-, o-C_6H_4$$
$$R' = CH_3, C_6H_5$$
$$R^2 = CH_3, C_6H_5, OC_2H_5$$

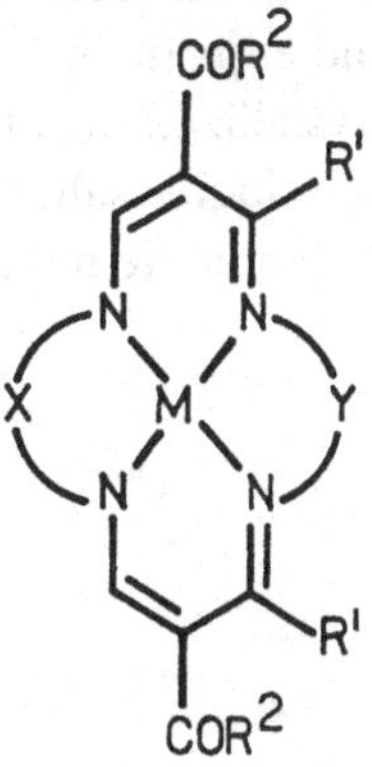

(74)   (75)

(76)   + excess $\xrightarrow[\text{tetrahydro-furan/ethanol}]{\text{OH}^-}$ (77)   [2.17]

recently, a refined synthetic procedure, illustrated by [2.17], for these systems has been developed (Melson & Funke, 1984). Mechanistic details of the synthesis were also investigated. Although aspects of this latter study are not completely straightforward, the results obtained are in accordance with the condensation proceeding via a multistep process in which there is initial monodentate coordination of 1,2-diaminoethane in axial positions of (76). Step 2 was proposed to involve deprotonation of a coordinated amine group of one of the 1,2-diaminoethane adducts followed by a slow rate-determining (*cis*) nucleophilic attack of the deprotonated amine at the carbon of a coordinated carbonyl group belonging to the in-plane ligand. Finally, it was suggested that a rapid ring-closing reaction occurs to yield the cyclic product (77).

The synthesis of the Ni(II) complex of the 13-membered (anionic) macrocycle (78) is also achieved using an *in situ* procedure (Cummings & Sievers, 1970) in which triethylenetetramine, acetic acid, acetylacetone, and nickel acetate are heated in water at the reflux. Addition of iodide ion and adjustment of the pH of the solution to approximately 10, leads to crystallization of the Ni(II) complex of the required cyclized product (78) as its iodide salt. The reaction type has been extended to include Cu(II) as the template metal (Martin, Wei & Cummings, 1972) and has also been

(78)

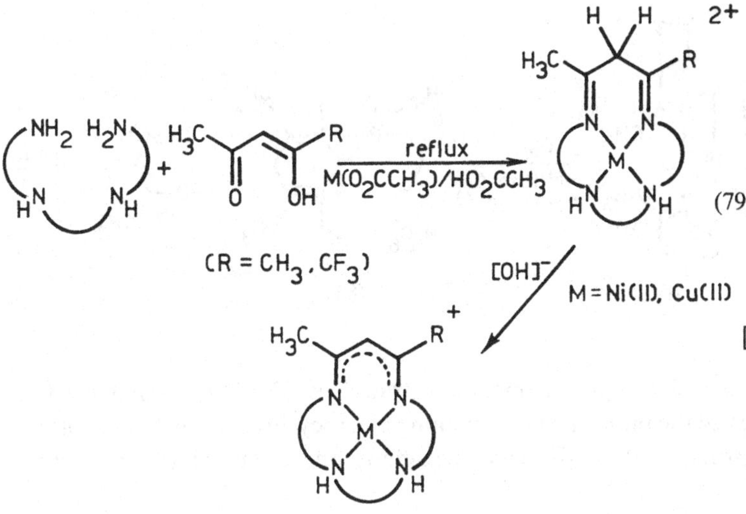

(79)

(R = CH₃, CF₃)

[2.18]

M = Ni(II), Cu(II)

(80)

adapted for the synthesis of analogous larger-ring complexes (80) (Martin & Cummings, 1973). In general terms the procedure may be represented by [2.18]. Protonated ligand complexes of type (79), incorporating the 14-membered ring macrocycle, have also been crystallized for both Ni(II) and Cu(II). In no cases have the respective ligands been isolated free of their metal ions. The relative acidity of the coordinated ligand in complexes of type (79) is markedly dependent on whether M = Ni(II) or Cu(II): the pK_a of the Cu(II) complex of the 14-membered ring is 9.3 and thus this complex is considerably less acidic than the corresponding Ni(II) species which has a pK_a of 6.45.

In an alternative synthetic procedure, the Ni(II) complex of the deprotonated form of (81) undergoes rearrangement on heating in water (at 100 °C) at pH 5 over 6 hours. Addition of iodide ion and adjustment of the pH to 10 results in precipitation of the corresponding macrocyclic complex (77) as its iodide salt. More recently, the related acid-catalyzed

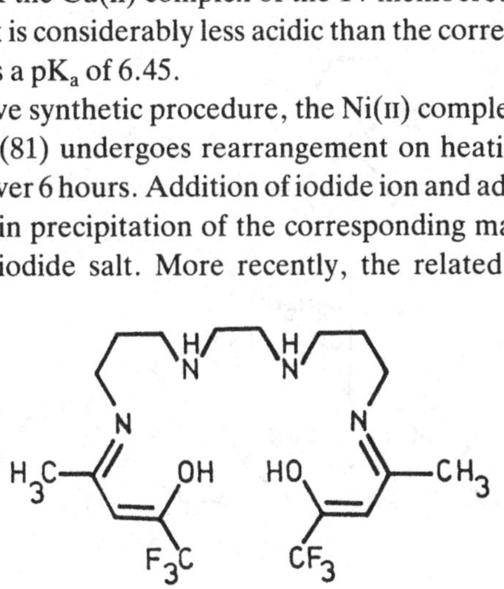

(81)

$$[2.19]$$

(82)

intramolecular rearrangement illustrated by [2.19] has been used to synthesize the phosphorus-containing macrocycle (82) which was isolated (from solution at pH 4) as its dihexafluorophosphate salt (Scanlon *et al.*, 1980).

Reactions of selected metal complexes of multidentate amines with formaldehyde and a range of carbon acids (such as nitroethane) have led to ring-closure reactions to yield a series of three-dimensional cage molecules (see Chapter 3). Condensations of this type may also be used to produce two-dimensional macrocycles (Comba *et al.*, 1986) – see [2.20]. In such cases, it appears that imine intermediates are initially produced by condensation of the amines with formaldehyde as in the Curtis reaction. This is followed by attack of the conjugate base of the carbon acid on an imine carbon. The resulting bound (new) carbon acid then reacts with a second imine in a *cis* site to yield chelate ring formation.

$$[2.20]$$

Studies involving 2,6-diacetylpyridine derivatives. Detailed studies by Nelson and coworkers supported by X-ray structural work by Drew and coworkers have elucidated many aspects of the template synthesis of metal complexes of a Schiff-base ligand series which includes the N_4-donor system (83) and the N_5-systems (84)–(86) (Nelson, 1980). Since

(83)

(84) $m = n = 2$ (15-membered)
(85) $m = 2$, $n = 3$ (16-membered)
(86) $m = 3$, $n = 2$ (17-membered)

such systems have been investigated so thoroughly and, because the syntheses appear typical of a range of other template reactions, it is appropriate to discuss the results obtained by these workers in some detail. A range of metal complexes of the above ligands has been obtained by a typical *in situ* procedure; namely, condensation of the appropriate triamine or tetramine with 2,6-diacetylpyridine in methanol or ethanol in the presence of a suitable metal ion. In the absence of the metal ion only viscous oils were obtained which were assumed to be the corresponding polymeric products.

It is probably of importance to the condensation that 2,6-diacetyl-pyridine incorporates a good donor (the pyridine nitrogen) between the

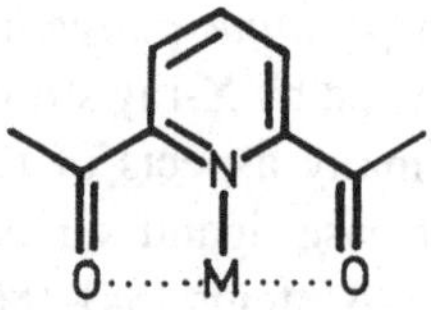

(87)

carbonyl functions since this will tend to ensure binding of the carbonyls (normally weak donors) to the template metal as in (87). If binding occurs, there is ample evidence from other studies that the carbonyl carbons will be activated towards nucleophilic attack.

Condensation of 2,6-diacetylpyridine with bis(3-aminopropane)amine in the presence of small ions such as Mn(II), Co(II), Ni(II) or Cu(II) readily leads to formation of the corresponding monomeric (14-membered) macrocyclic complexes of ligand (83). However, when the large Ag(I) ion is used as the template, then a dimetallic complex of a 28-membered macrocycle of type (88) is produced. This example illustrates well the importance of metal-ion size in promoting template reactions.

The role of the metal ion in the template formation of complexes of the N_5-macrocycles (84)–(86) has been studied in considerable detail. Thus, the following metal ions were investigated for their effectiveness in promoting cyclization: Mg(II), Ca(II), Sr(II), Ba(II), Al(III), La(III), Mn(II), Fe(II), Fe(III), Co(II), Ni(II), Cu(II), Ag(I), Zn(II), Cd(II), Hg(II), Sn(IV), and Pb(II). The radii of these ions range from 0.53 Å (Al^{3+}) to 1.36 Å (Ba^{2+}). Not all of these metals acted as templates for obtaining (84)–(86): some were ineffectual templates for all the rings while others were only effective for certain rings. One factor which appears to

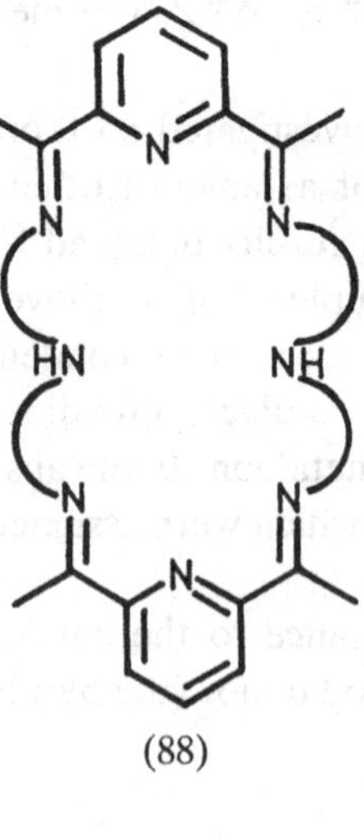

(88)

influence the ability of a particular ion to act as a template is its compatibility or otherwise for the macrocyclic cavity formed in the product. For example, to produce the largest (17-membered) ring (86), a metal ion of radius greater than ~0.8 Å appears to be required. However, it was also noted that the smaller rings sometimes form in the presence of the larger metals. For some systems in which this occurs there is structural evidence that the complexed metal ion is displaced from the plane of the macrocyclic ring such that its effective template radius is smaller than its true radius.

In view of the points just discussed, it is perhaps not surprising that the relatively small Ni(II) or Cu(II) ions do not act as templates for the formation of (84)–(86). However, another factor besides size is probably important in these cases. Transition metal ions such as Ni(II) and Cu(II) generally show a preference for stereochemistries in which the bonding orbitals are orthogonal. Thus such ions will tend not to stabilize the pentagonal-planar coordination of (83)–(85) which, from X-ray studies, seems to be the preferred ligand geometry in a range of complexes of non-transition metal ions. In the latter, besides the planar coordination of the N_5-ligand, axial ligands are often also present. Thus the following overall metal-coordination geometries have been observed: five-coordinate (distorted pentagonal plane), six-coordinate (distorted pentagonal pyramid) and seven-coordinate (pentagonal bipyramidal).

It is of synthetic significance that metal exchange reactions involving complexes of (84)–(86) have also been investigated. For example, if an alcohol solution of $[AgL]_2(ClO_4)_2$ (L = 84) is treated with an alkali halide, the silver is removed as the corresponding insoluble halide and decomposition of the free macrocycle occurs. However, if Ni(II) is present in the solution then a metal exchange reaction occurs. The product is a Ni(II) complex containing the more flexible ligand derivative (89) formed by addition of an alcohol molecule across one of the imine

(89)

(90)

bonds of (85). If the reaction is performed in water, full hydrolysis of one imine is observed [to yield an octahedral complex of the corresponding open-chain derivative (90)]. Such ligand reactions appear to be a reflection of the strong preference of Ni(II) for an octahedral coordination shell. However, for a range of other less sterically-demanding metal ions the opposite process can be considered to occur: namely, the stereochemical dictates of the N_5-ligand impose an unusual (for example, pentagonal pyramid or pentagonal bipyramid) geometry on the respective metal ions.

When the nickel exchange reaction was attempted using the Ag(I) complex of the 17-membered macrocycle (86), then the Ni(II) complex which resulted was found to contain this macrocycle intact. In this case, the larger ring size of the ligand [relative to the 16-membered ring (85)] allows sufficient flexibility for the macrocycle to occupy five positions around an octahedron. Thus, in this case, the more stable octahedral geometry is attained without the need for solvolysis of an imine linkage.

A related study involving a 2,5-diformylfuran derivative. In an extension of the above studies, the template synthesis of (91), derived from a 2:2 condensation of 2,5-diformylfuran with 1,3-diaminopropane has been investigated (Drew, Esho & Nelson, 1983). The larger alkaline earth metals, Ca(II), Sr(II) and Ba(II), act as templates for this ring but Mg(II) does not. Pb(II) has also been shown to promote the formation of (91). These results confirm the conclusions from the previous studies, namely, that the effectiveness of these ions as templates is often associated with the product complexes attaining reasonable thermodynamic stabilities which, in turn, is related to the match of the macrocyclic cavity for the radii of the ions. The absence of strong stereochemical preferences on the part of these 'hard' metal ions is also likely to be of importance to their successful role as templates for this ligand system (which exhibits a

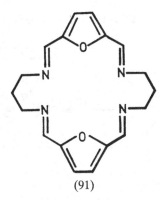

(91)

tendency for the donor atoms to be orientated in a hexagonal-planar array).

Although the alkaline earth complexes of (91) are moderately stable, the Ba^{2+} complex is still able to undergo exchange reactions with Co(II) and Cu(II). These latter metal ions yield products exhibiting a 2:1 metal to ligand ratio. In the formation of these latter complexes, a template reaction involving a larger alkaline earth ion is followed by a transition metal exchange to yield a binuclear complex which is not readily accessible by other means (Drew, Esho, Lavery & Nelson, 1984). This reaction type shows much potential since such binuclear complexes are of considerable interest for use as models for several classes of biological molecules which incorporate interacting metal-ion sites.

Summary of typical ring-closing reactions

It is appropriate at this point to summarize (Table 2.1) the ring-closing reactions met so far in this chapter and which have been claimed to occur via template means. Only reactions which bridge *cis* coordination positions have been included. It is stressed that the aim of the table is to present, in schematic form, some typical ring-closing reactions for which reactants and products are identified; the initial coordination of a functional group is not meant to imply that it necessarily remains coordinated throughout the course of a particular bridging reaction. Similarly, most of these reactions are quite sensitive to the chosen reaction conditions and are influenced by such factors as: the nature of the solvent, the pH, the order of addition of reactants, the temperature, the time used for the reaction and/or the need (or otherwise) to protect the system from the atmosphere. As discussed already, particular reactions will also be quite dependent on the nature of the template metal chosen for the reaction.

Table 2.1 Summary of *in situ* ring-closing reactions

<u>REAGENTS</u>   <u>PRODUCTS</u>

Table 2.1 *continued*

REAGENTS	PRODUCTS

2.4   Further considerations

Modification of the preformed macrocycle

Following the generation of a macrocyclic ring by a template reaction it has often been the practice to modify the ring by further reaction. Examples of such modification are given by the formation of the derivative complexes (71) and (72) discussed in Section 2.3. Other examples are described in subsequent chapters – in particular, ligand redox reactions have been widely used and a range of such reactions is presented in Chapter 8.

The ligand reaction step may occur either with the template metal still intact or may take place after removal of the metal ion from the ring. As already mentioned, many of the Schiff-base macrocycles are unstable in the absence of a coordinated metal ion. However, for such systems, it has often been possible to hydrogenate the coordinated imine functions directly. The resulting saturated ligands will not be subject to the hydrolytic degradation which occurs for the imine precursors in the absence of their metal ion.

Typically, the imine linkages can be reduced chemically (for example, using sodium borohydride in methanol), catalytically (for example, using H_2 in the presence of a catalyst such as Pd/C) or electrochemically. The reduced ligands can normally be removed from their respective metal ions using one of a number of procedures.

Generation of free macrocycles from their metal complexes

The removal of a macrocycle from its coordinated metal ion is a frequent procedure in macrocyclic ligand synthesis – conditions for inducing such demetallations are now briefly discussed:

(i) The addition of excess acid may lead to demetallation of the complex of an amine-containing macrocycle. For chemically labile systems, the acid will protonate the amine functions as they dissociate from the metal ion and thus 'scavenge' the macrocycle as its N-protonated form.

(ii) Demetallation may be induced by addition of a strongly competing ligand to a solution of the macrocyclic complex [the cyanide ion or the analytical reagent, ethylenediaminetetraacetate (edta) are frequently used]. In some cases, for example when sulfide or hydroxide ion is employed as the scavenging ligand, the metal may be removed as an insoluble precipitate (the metal sulfide or hydroxide) leaving the metal-free macrocycle in the supernatant liquid.

(iii) In special cases, when the template ion is weakly coordinated, demetallation may be induced simply by dissolution of the complex in a coordinating solvent in which the free macrocycle has poor solubility (Lindoy & Busch, 1969). Indeed, in particular instances, free Schiff-base macrocycles have been isolated directly from *in situ* reaction mixtures in which a weakly coordinating metal ion is used as the template (Green, Smith & Tasker, 1971; Black, Bos, Vanderzalm & Wong, 1979). Related behaviour in which dimethyltin(IV) directs the course of a cyclization without tin being itself incorporated in the macrocyclic product has been investigated. The general phenomenon has been termed the 'transient template effect' (Constable *et al.*, 1985). The origins of the effect in this latter case appear to lie in the ability of dimethyltin(IV) to act as a template for the immediate open-chain precursor of the cyclic system. However, this ion appears too large for the cavity of the macrocyclic product formed. Under the conditions employed for the ring-closing step, the tin precipitates as its insoluble oxide leaving the metal-free macrocycle in solution. The latter may then be isolated as its hydrogen hexafluorophosphate salt.

(iv) For some systems it has been found necessary to perform a redox reaction (usually a reduction) on the complexed metal before it may be removed from the macrocycle. This is often the case when the metal ion in its original oxidation state is kinetically inert [for example, Co(III) or Cr(III)]. On reduction, the lability of such ions is normally increased and it becomes feasible to use one of the procedures just discussed to remove the ion from its macrocyclic complex. For some systems, reduction of the metal ion to a low oxidation state [for example, to the M(0) state] leads to its spontaneous dissociation from the macrocycle.

2.5 Concluding remarks

Resulting from the widespread interest in macrocyclic ligand chemistry, an impressive array of synthetic procedures for macrocyclic systems has been developed. In spite of this, the synthesis of a new ring system frequently turns out to be far from trivial. Indeed, synthetic macrocyclic chemistry is often very challenging and, as well as a thorough knowledge of any metal-ion chemistry involved, skill in the subtle 'art' of organic synthesis is also a valuable asset for those involved in the preparation of new cyclic systems.

Besides the simple-ring ligands discussed in this and the previous chapter, a number of other macrocyclic types have been reported. The next three chapters survey the chemistry (including some synthetic details) of these further categories of macrocyclic ligands. Thus, in Chapter 3 there is presented a discussion of 'structurally developed' cyclic systems. Such systems include: rings to which 'arms' incorporating additional donor atoms have been appended, macrocycles which are able to incorporate more than one metal ion, interlocking macrocyclic systems, three-dimensional macrocycles which incorporate a void adjacent to the metal binding site, and cage macrocycles which are capable of tightly encapsulating a metal ion. Chapter 4 is devoted to the large category of cyclic systems incorporating polyether donor sets while Chapter 5 describes 'host' systems which are capable of forming inclusion complexes with a variety of non-metal 'guest' species.

3

Macrocyclic systems – some further categories

3.1 Preliminary remarks

There has been a steady tendency in the field of macrocyclic chemistry to undertake the synthesis of new ring systems which exhibit ever-increasing structural diversity. Such investigations have resulted in several new categories of cyclic systems being developed. In this chapter, a number of such 'derivative' macrocyclic systems are discussed. For each category, a range of representative complexes is described, together with a discussion of selected properties for each ring system. Thus, each chapter section gives a 'cameo' view of these ligand categories. Collectively, the discussions also provide a perspective of some of the important advances in the continuing development of new macrocyclic systems.

3.2 Macrocycles with pendant functional groups

There has been considerable activity concerned with the synthesis of cyclic systems related to the monocyclic rings discussed in Chapters 1 and 2 but which also contain appended side chains incorporating additional donor functions. Such products have been obtained both from structural modification of selected simple rings as well as via synthetic procedures designed to produce the required macrocycle directly from non-cyclic precursors. Products of this type are structurally related to both the simple macrocyclic as well as the open-chain ligand categories. For example, many of the complexes exhibit the kinetic inertness which is typical of cyclic systems while still showing some of the coordination flexibility frequently associated with open-chain ligands.

Many ligands of this category offer the prospect of inducing axial metal-ion coordination even for those cases where the pendant arms incorporate weak donor functions. Coordination will be enhanced simply because the donors are held near to the metal and hence their 'effective' concentration with respect to the metal is greater than their 'true' concentration (with respect to the bulk solution). Proximity effects of this type have been well-documented in other systems.

Some representative systems

Simple rings with flexible pendant 'arms'. Structure (92; X = NH, S) represents two small-ring examples of the pendant arm class. These potentially quinquedentate ligands form Co(III) complexes of type [Co(ligand)Cl]$^{2+}$ in which the respective ligands have their five hetero atoms bound to the cobalt such that this ion achieves its normal coordination number of six (Gahan, Lawrance & Sargeson, 1982). Ligand (92; X = NH) can be considered to be an analogue of the linear pentaamine, 3,6,9-triazaundecane-1,11-diamine (93), in which a $-(CH_2)_2-$ group links the nitrogens of (93) at positions 3 and 9. However, for (93),

X = NH, S

(92)   (93)

there are four geometrical isomers which in principle may occur when this ligand coordinates around five positions of an octahedral metal ion. The presence of the additional bridge in the macrocyclic derivative restricts the number of coordination modes such that only two isomers, (94) and (95), are possible. As occurs with this system, flexible macrocyclic ligands usually exhibit a reduced number of possible coordination modes when compared with their open-chain analogues.

(94)   (95)

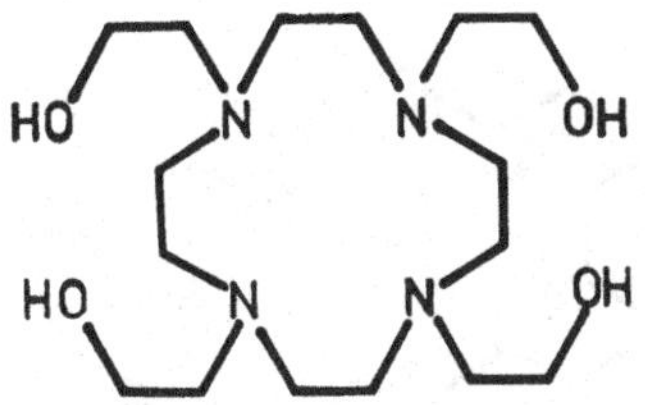

(96)

The cyclic tetraalcohol (96) uses a varying number of the pendant alcohol functions for complexation to alkali cations (Buøen, Dale, Groth & Krane, 1982) leading to complexes of lithium, sodium, and potassium in which penta-, hepta-, and octa-coordination occurs, respectively. The corresponding tetraalcohol derivative of cyclam has also been synthesized (Hay & Clark, 1984; Madeyski, Michael & Hancock, 1984) and has been demonstrated to form complexes with a number of transition and non-transition metal ions. Once again some (or all) of the alcohol groups remain uncoordinated in these complexes. Further, non-coordination of the hydroxyethyl arms in the complex [MnLCl][BF$_4$]·4H$_2$O (where L = 97) has also been demonstrated in the solid using X-ray diffraction.

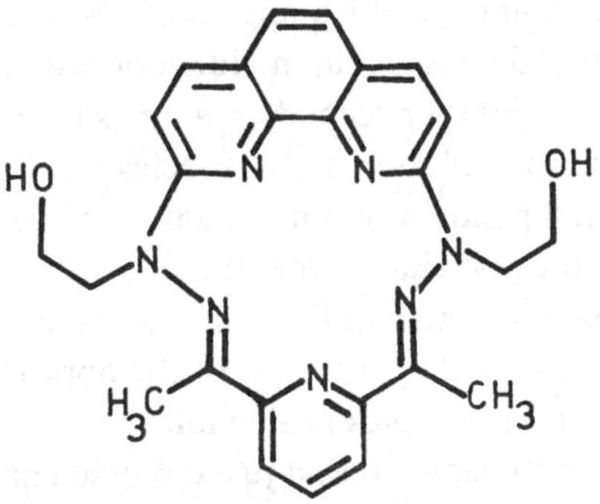

(97)

The varying participation of the alcohol functions in metal coordination in these complexes is undoubtedly a reflection of both the moderate donor capacity of alcohol oxygens and the different coordination preferences of the respective metal ions involved.

The product from heating cyclam directly in acrylonitrile is the tetranitrile derivative (98, X = CN) which does not use any of its nitrile groups for coordination to Ni(II) or Cu(II) (Wainwright, 1980; Hay & Bembi, 1982; Freeman, Barefield & Van Derveer, 1984). This behaviour con-

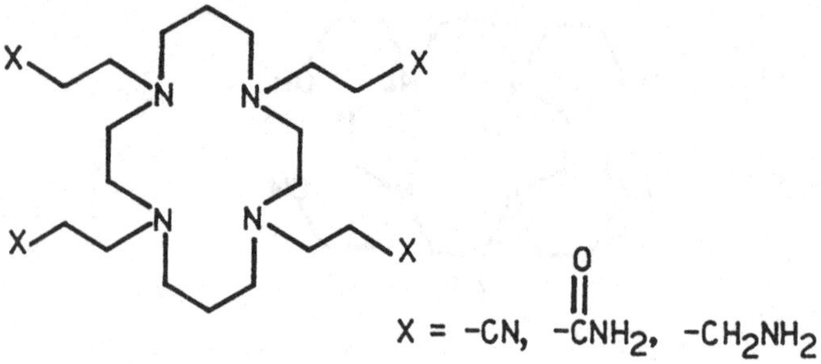

X = –CN, –C(O)NH$_2$, –CH$_2$NH$_2$

(98)

trasts with that of the corresponding amide derivative [(98), X = C(O)NH$_2$] which yields two isomeric Ni(II) species; for one of these, the four nitrogens of the ring and two pendant amide groups were shown by X-ray diffraction to bind to the metal. Overall, an octahedral coordination geometry is achieved. The second isomer was assigned a five-coordinate structure in which only one of the amide groups is bound to the Ni(II) ion (Freeman, Barefield & Van Derveer, 1984).

Reduction of (98, X = CN) with sodium and ethanol in toluene generates the octaamine (98, X = CH$_2$NH$_2$) (Wainwright, 1983). The addition of nickel ion to an ethanol solution of this derivative at neutral pH yields a high-spin complex which appears to be five-coordinate and contains one coordinated pendant amine function with the coordination sphere being completed by four secondary amines from the macrocyclic ring. Addition of acid to a solution of this species results in decomplexation of the coordinated pendant donor via amine protonation (together with protonation of the remaining free arms). Ni(II) complexes of the resulting tetra-protonated macrocycle have been isolated – the protonation process may be reversed by addition of the appropriate amount of base to a suspension of the complex in alcohol.

A related pH-induced change to that just discussed had been reported earlier for the nickel complex of (99) which undergoes a colour change

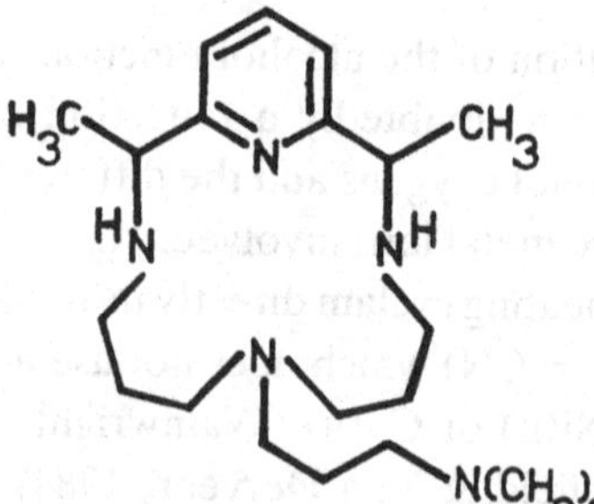

(99)

from yellow (low spin) to blue (high spin) when an acid solution of the complex is made alkaline (Lotz & Kaden, 1978). Under acid conditions the dimethylamine group is protonated and the nickel ion is then surrounded solely by the four donor atoms of the macrocyclic ring to give a square-planar species. Deprotonation of the dimethylamine function in basic solution releases an additional donor group which is then able to coordinate in an axial position and a high-spin nickel species results. The acid–base behaviour of these branched-chain systems serves to emphasize the inertness towards dissociation from the metal ion of the nitrogens of the macrocyclic rings relative to those attached to the pendant arms.

The synthesis of (99) provides an example in which the pendant group is incorporated in one of the reactants before the cyclization reaction occurs. The synthesis proceeds via two steps. The first is the template condensation shown in [3.1] while the second involves the catalytic hydrogenation of the imine functions of (100) to yield (99). A similar reaction sequence starting from (101) leads to the corresponding species containing an alcohol pendant group. Both of the hydrogenated ligand species can be obtained free of Ni(II) by treatment of the respective complexes with excess NaCN. The free ligands readily react with a number of other metal ions to yield the corresponding complexes.

A variety of other mono-*N*-substituted tetraazamacrocycles have been prepared (Lotz & Kaden, 1978; Hediger & Kaden, 1978; Keypour &

(101)

Stotter, 1979) in which single pendant arms incorporate functional groups such as $-CH_2CONH_2$ and $-CH_2CH_2CH_2NH_2$. The metal derivatives of these systems show a superficial resemblance to particular biochemical systems such as haemoglobin, in which the metal ion is surrounded by four nitrogen atoms in a square-planar fashion with a fifth donor, attached to another part of the molecule, occupying an axial coordination site.

Rings with rigid pendant 'arms'. The interesting ligand tris(2,3-dihydroxybenzoyl)1,5,9-triazacyclotridecane (102), was synthesized specifically to act as a reagent for Fe(III) (Weitl & Raymond, 1979). It is structurally quite closely related to enterobactin (103), the natural molecule used by *E. coli* to transport Fe(III) through its cell walls. The protonation and complexation equilibria of Fe(III) with (102) have been

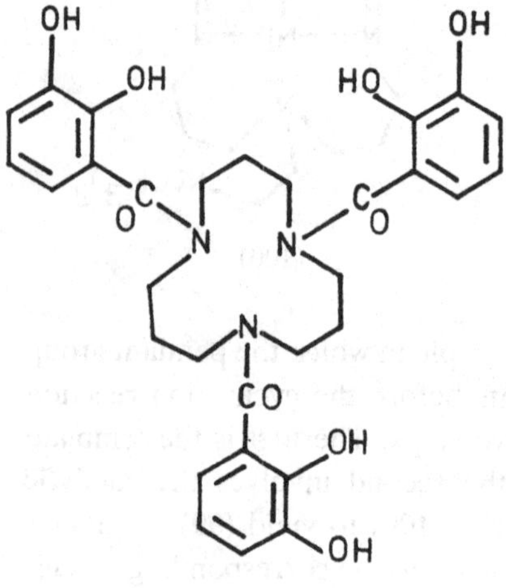

(102)

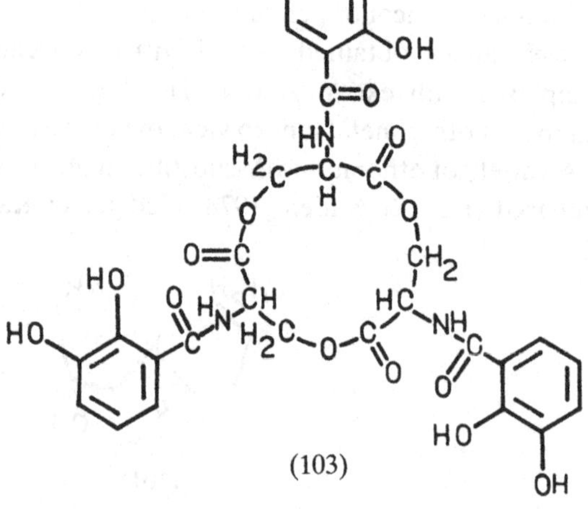

(103)

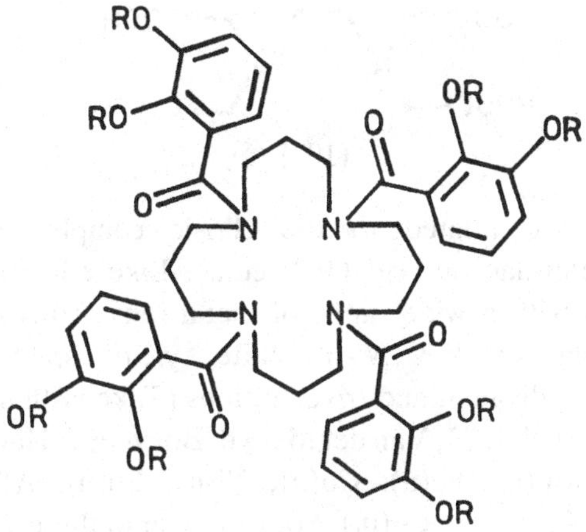

(104)

investigated (Harris, Raymond & Weitl, 1981). Under strongly alkaline conditions the ligand coordinates to Fe(III) via all six phenolic oxygens to yield an extremely strong complex. Indeed, (102) is able to strip iron from the Fe(III) protein, transferrin. As the pH is lowered, protonated ligand complexes of (102) are formed in which the ligand is bound less strongly to the Fe(III). In a similar manner, related larger-ring species (104) have been designed for the specific complexation of (the larger) actinide ions (Weitl, Raymond, Smith & Howard, 1978).

The presence of the cyclic backbone in ligands of this type makes a substantial contribution to their metal-ion complexing ability even though coordination involves donors which are not directly incorporated in the ring fragment. The origins of the enhanced stability of the metal-containing species may be considered to reflect the operation of an 'indirect' macrocyclic effect (see Chapter 6) in these systems.

Complexone-like derivatives. Cyclic systems containing attached carboxylic acid functions such as (105) and (106) have been synthesized.

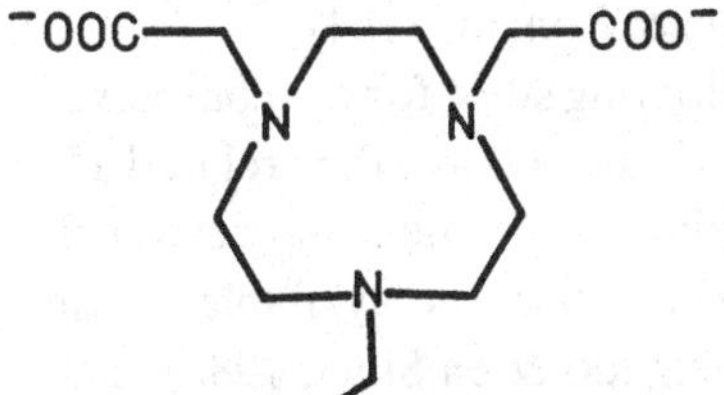

(105) 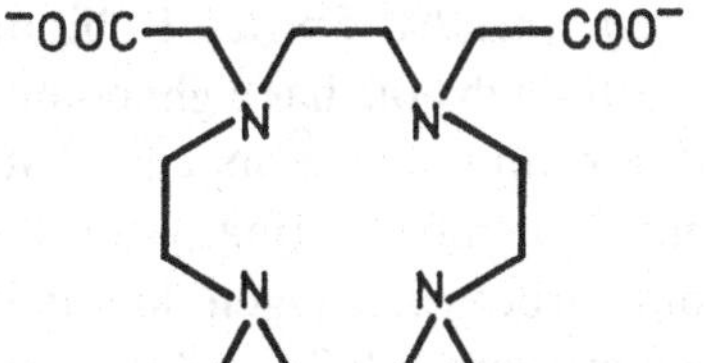   (106)

HOOC⏤⟍   ⟋⏤COOH
        N   N
HOOC⏤⟋   ⟍⏤COOH

(107)

These systems are related to the classic complexones such as ethylenediaminetetraacetic acid (107, edta). Like edta, these ligands form complexes with a wide range of metal ions. Thus ligand (105), 1,4,7-triazacyclononane-N,N',N''-triacetate, yields stable complexes with the following divalent and trivalent ions (Takahashi & Takamoto, 1977, Wieghardt *et al*, 1982; Van der Merwe, Boeyens & Hancock, 1983): Mg(II), Ca(II), Mn(II), Fe(II), Co(II), Ni(II), Cu(II), Al(III), Cr(III), Mn(III), Fe(III), Ni(III), and Co(III). For a number of these complexes the sexadentate coordination of (105) has been confirmed by X-ray diffraction studies; coordination geometries approximating both octahedral (108) and trigonal prismatic (109) have been established for specific complexes.

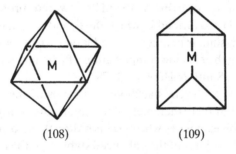

(108)   (109)

As with (105), ligand (106, dota) also forms strong complexes with a range of both non-transition and transition metal ions (Stetter & Frank, 1976; Delgado & da Silva, 1982; Spirlet, Rebizant, Desreux & Loncin, 1984) which are often more stable than the corresponding edta complexes. In particular, the calcium complex shows extremely high stability and very stable complexes are also formed with the trivalent lanthanides (Desreux, 1980; Spirlet, Rebizant, Desreux & Loncin, 1984).

Although dota has eight potential coordinating sites, for the complexes of transition ions it appears that this ligand (as well as other related 15- and 16-membered ring systems) may coordinate in some instances via all four carboxylic oxygen donors but use only two of the available amine donor atoms (Hafliger & Kaden, 1979; Delgado & da Silva, 1982). The kinetics of formation of dota and the analogous tetraacetate derivative of cyclam (teta) has been investigated for a range of both non-transition and

(divalent) transition ions (Kasprzyk & Wilkins, 1982); the monoproton-
ated (LH^{3-}) form of the ligand was characterized as the reacting species
in each case.

The diacetate derivative (110) has also been shown to form complexes
with a wide range of metal ions (Chang & Rowland, 1983). Once again,
this is an excellent ligand for the lanthanide ions. With Cu(II), (110) yields
a 1:1 complex in which the copper is coordinated by two ring nitrogens
and two acetate oxygens in a *trans* square-planar arrangement. In addition, two ether oxygen donors are weakly bonded to the copper on
either side of the square plane such that a distorted octahedral structure
results (Uechi *et al.*, 1982).

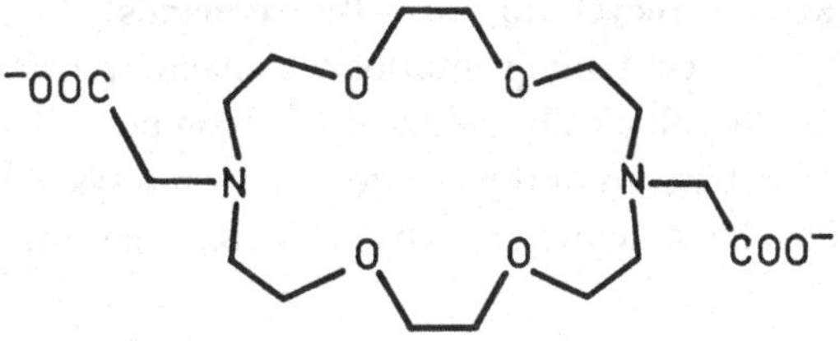

(110)

A 'picket fence' porphyrin system. Finally, examples of pendant donor
groups attached to a porphyrin ring are known. A novel example of such
a molecule is *meso-α,α,α,α-*tetrakis(*o*-nicotinamidophenyl)porphyrin
(111) which is capable of binding two metal ions such that each has a
square-planar environment with the square planes oriented coaxially to
each other (Gunter *et al.*, 1980). When a kinetically-inert metal ion such

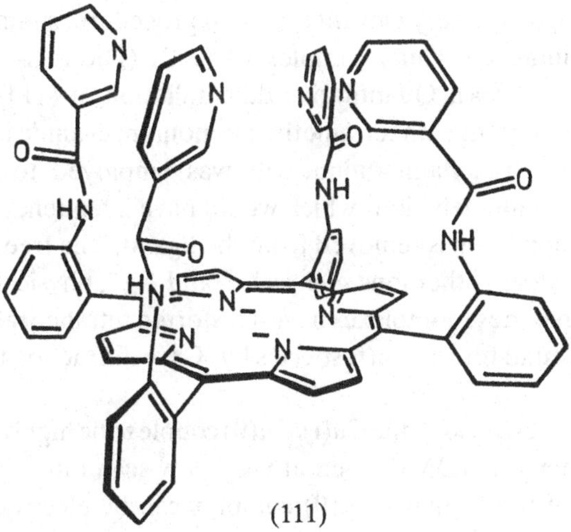

(111)

as Ru(II) is reacted with this ligand, coordination of the four pyridyl groups occurs to the ruthenium such that this ion locks the nicotinic 'pickets' into place and inhibits the ligand isomerization which has been shown to occur in the presence of more labile metal ions. The reaction of the Ru(II) species with other transition metal ions leads to the formation of a series of mixed-metal complexes in which the second metal is contained in the cavity of the porphyrin ring. Ligands of this type hold two metal centres in close proximity and in this respect are related to a range of other large ring macrocycles which are capable of incorporating two metal ions within the ring. Examples of this latter category are discussed in Section 3.3.

3.3   Interlocked macrocyclic ligands – the catenands

Cyclic ligands containing interlocked macrocyclic rings are a further category of topologically-unique cyclic compounds. Following the synthesis by direct means of the interlocked ligand system (112), for which each ring has the structure given by (113), Sauvage and coworkers

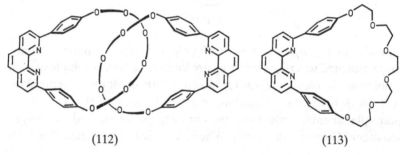

(112)   (113)

developed a simple and very elegant template procedure, summarized by [3.2], for obtaining the Cu(I) complex of (112) (Dietrich-Buchecker, Sauvage & Kern, 1984). Quantitative demetallation of (114) was then performed by treatment with tetramethylammonium cyanide in acetonitrile/water. The tetralkylammonium salt was employed to avoid the introduction of alkali metal ions which would have a tendency to coordinate once the copper ion is removed from the ligand. The free ligand has been used to complex other ions such as Li^+ and Ag^+; physical measurements suggest that these complexes have a distorted tetrahedral geometry related to that found for the Cu(I) species by X-ray diffraction (Cesario *et al.*, 1985).

Cyclic voltammetry shows the Cu(I)/Cu(0) couple to be highly reversible in dimethylformamide (DMF) even at very slow scan rates. Thus, dark blue solutions of the formally Cu(0) complex can be electrochemically

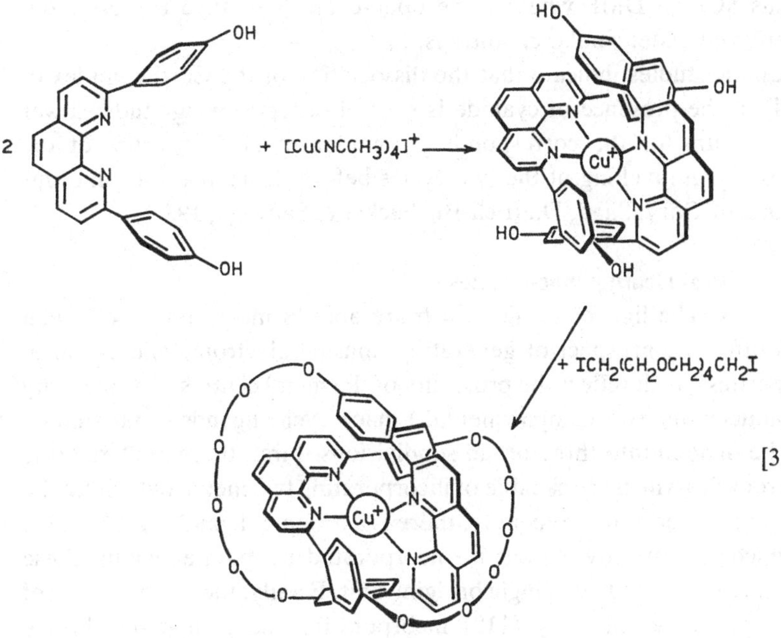

(114)

generated at -1.67 V versus SCE in this solvent. The unusual stability of this Cu(0) species appears to be a direct consequence of the presence of interlocking rings since parallel studies involving the similar bis-ligand complex of the corresponding open-chain ligand (115) showed that this latter species lacks the stability of the interlocked system. Hence, the interlocked system is not electrochemically demetallated at -1.8 V

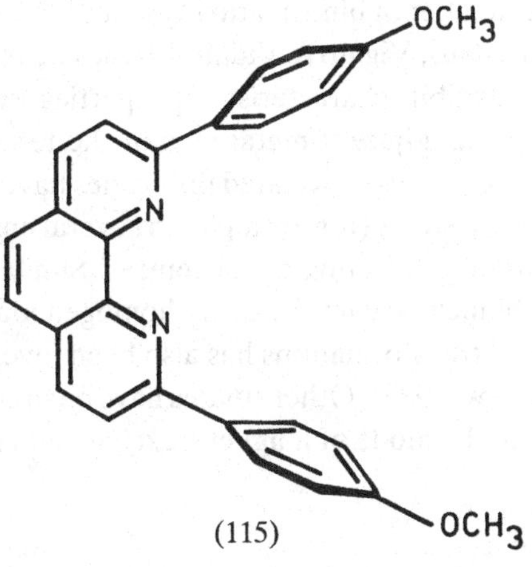

(115)

versus SCE in DMF whereas the open-chain derivative is readily de-complexed under similar conditions.

Kinetic studies indicate that the dissociation of the Cu(I) complex of (114) in the presence of cyanide is several orders of magnitude slower than occurs for the corresponding complex of (115) because of the necessary unravelling of the two cycles before demetallation can occur (Albrecht-Gary, Saad, Dietrich-Buchecker & Sauvage, 1985).

3.4   Binucleating macrocycles

Cyclic ligand species which are able to incorporate two metal ions offer the prospect of generating unusual electronic and chemical properties which reflect the proximity of the metal centres. The majority of binucleating (or 'compartmental') macrocyclic ligands so far studied can be divided into three broad subdivisions. First, there are large ring macrocycles which are capable of incorporating two metal ions within the great ring – see (116). Secondly, there is a group of ligands of type (117) in which the respective metals are incorporated in separate rings but these rings are connected by a single bridging unit. Finally, there are systems of the type represented by (118) incorporating more than one bridge between two ring systems.

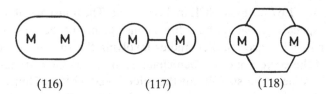

(116)   (117)   (118)

A considerable number of binucleating systems have now been synthesized (Fenton, Casellato, Vigato & Vidali, 1982). The resulting dimetallic complexes often exhibit characteristic properties such as magnetic exchange between the adjacent metal ions or the tendency to undergo multi-electron redox reactions. Related properties have been observed in a number of natural systems (for example, in several enzymes) which are known to incorporate interacting metal centres. Similarly, the potential use of synthetic binuclear complexes as homogeneous catalysts for a number of chemical transformations has also been investigated (Nelson, Esho, Lavery & Drew, 1983). Other studies have been concerned primarily with the study of the mode of magnetic exchange in complexes of this type.

Large-ring macrocycles incorporating only metal ions

Two early examples of the large-ring category of binucleating macrocyclic ligands are (119) (Travis & Busch, 1970) and (120) (Pilkington & Robson, 1970).

Ligand (119) yields nickel complexes of type $[Ni_2L](BF_4)_4$ (low spin) and $[Ni_2L(NCS)_4].2H_2O$ (high spin). For each of these complexes, physical studies indicate that the macrocycle circumscribes the two nickel ions such that each ion is surrounded by four sulfur donors in a planar array.

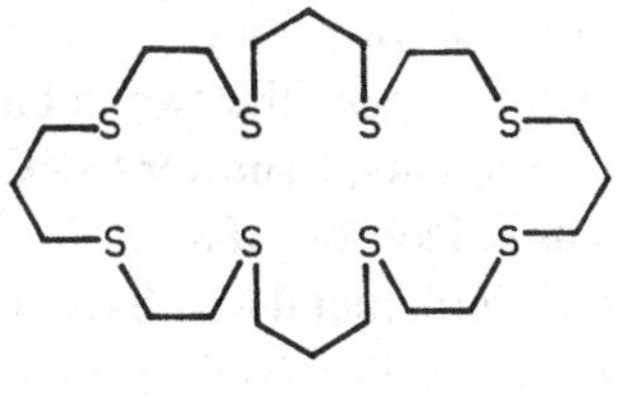

(119) 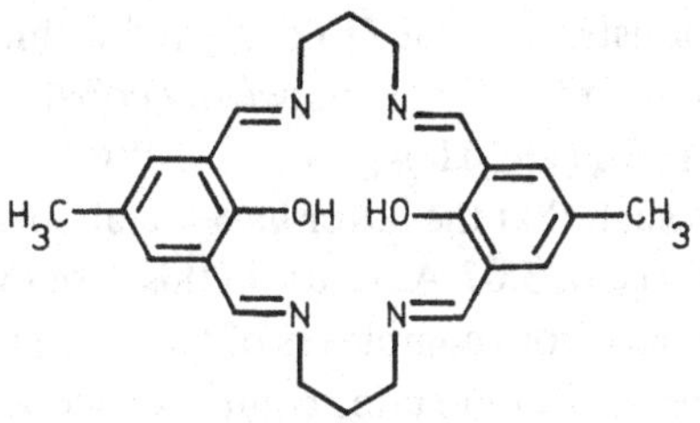   (120)

The metal-ion chemistry of (120) has been studied in more detail. This ligand was the first example of a binucleating macrocycle which shares donor atoms between metal-containing sites. On coordination, (120) loses its phenolic protons to give complexes of type (121, R = H) in which the phenoxide oxygens bridge between metal ions. In the majority of such complexes, monodentate ligands occupy axial positions on the metal ions which tend to have either five- or six-coordinate geometries (Hoskins, Robson & Williams, 1976). For example, a series of complexes of type M_2LCl_2. (solvent) where L is the dianionic form of (120) and M = Mn(II),

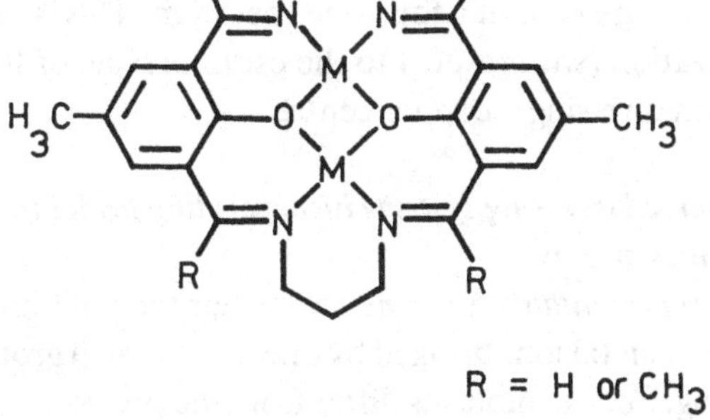

R = H or CH$_3$

(121)

Fe(II), Co(II), Ni(II), Cu(II), or Zn(II) have been isolated. On the basis of spectral and magnetic data, the complexes were assigned structures in which the cations are in approximately square-pyramidal environments. The X-ray structure of $Cu_2LCl_2.6H_2O$ confirmed a geometry of this type with a chloride ion occupying an axial position on each copper. The structure of $Co_2LBr_2.CH_3OH$ is similar and has bromides occupying the axial positions. The Co–Co bond distance in this complex is 3.16 Å (Hoskins & Williams, 1975a) with the Co(II) ions being positioned 0.3 Å out of the respective N_2O_2-planes. In an extension of this work, it was demonstrated that *O,O′*-diethyl dithiophosphate can use its two sulfur donors to bridge between cobalt centres to yield the complex $Co_2L[S_2P(OC_2H_5)_2]_2$. Each $S_2P(OC_2H_5)_2^-$ ligand spans the two cobalt ions such that the latter have octahedral environments; a shorter Co–Co distance of 3.07 Å occurs in this case (Williams & Robson, 1981).

Apart from complexes of type (121) in which both metal ions have the same oxidation state, complexes containing metals in mixed oxidation states are also known. In particular, species incorporating Co(II)/Co(III) (Hoskins & Williams, 1975b; Hoskins, Robson & Williams, 1976) and Cu(I)/Cu(II) (Addison, 1976; Gagne, Koval & Smith, 1977) have been prepared. The X-ray structure of the mixed cobalt complex indicates that both cobalt ions have octahedral geometries and that the Co–Co distance is 3.13 Å. For the copper complex [121; M = Cu(I), Cu(II)], the solution esr spectrum contains seven lines and is hence consistent with the odd electron [arising from the d^9 Cu(II) ion] being delocalized over both copper centres ($I = \frac{3}{2}$). However, when R = CH_3, the complex exhibits a four-line esr spectrum in solution. Apparently, methyl substitution alters the conformational properties of the macrocycle sufficiently to inhibit thermal electron transfer between the copper ions on the esr time scale at temperatures for which it occurs when R = H. In accordance with this interpretation, frozen solutions of the complex with R = H at liquid nitrogen temperature exhibit a four-line spectrum. This is thus consistent with the localization (with respect to the esr time scale of 10^{-8}–10^{-4} s) of the odd electron on a single copper centre.

Dimetallic large-ring systems incorporating bridging small molecules or ions

Some representative systems. A further type of binuclear species incorporates two metal ions bridged by one or two small groups within the macrocyclic ring. Such complexes differ from the previous type in that the bridges between the metal ions do not involve donor groups which are directly attached to the macrocycle – see (122) and (123). A variety of

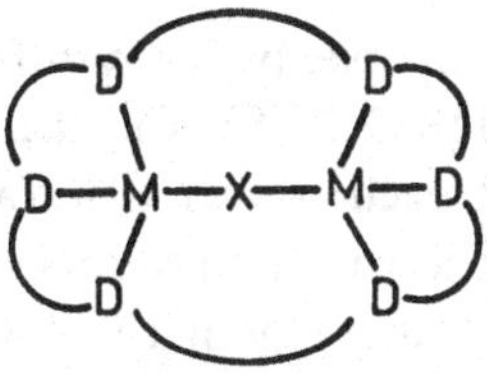

(122)

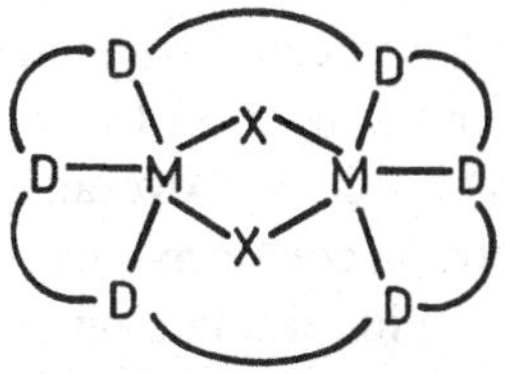

D = donor atom
X = bridging group

(123)

such complexes have been prepared (Nelson, 1982); examples of macrocycles (derived from 2,6-diacetylpyridine) involved are given by (124)–(126). Similar two-site macrocycles formed by condensation of 2,6-diformylpyridine or 2,5-diformylfuran have also been studied. These ligands vary in ring size from 18 to 30 members, and, when coordinated to two metal ions, are able to act as receptors for one or two bridging groups such as OH^-, OR^-, halide ion, N_3^-, NCS^-, $NCSe^-$, pyrazolate, imidazolate, or pyrazine. The smaller rings appear to accommodate only single-atom bridges (for example, OH^- or OR^-) while the larger rings may incorporate larger multi-atom bridges between the metal centres. When imidazolate is the bridging ion the complexes exhibit a structural resemblance to the active site of bovine erythrocyte superoxide dismutase. This enzyme has been shown by X-ray diffraction to contain a $Cu(II)\cdots Zn(II)$ unit which is bridged by an imidazolate group (Richardson, Thomas, Rubin & Richardson, 1975).

Particular attention has been given to the di-$Cu(II)$ complexes of ligands in the series (124)–(126); magnetic susceptibility and esr measurements have identified the presence of magnetic exchange in several such complexes and this exchange persists even when the copper ions are separated by a bridging ion as large as imidazolate. The structures of these

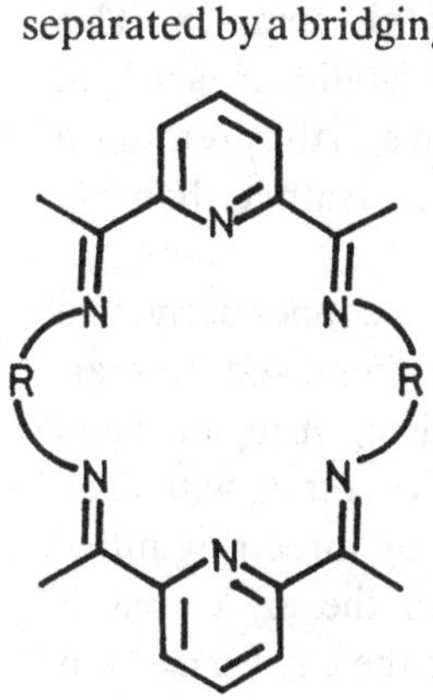

$R = -(CH_2)_2-$ (124)

$R = -(CH_2)_2-X-(CH_2)_2-, X = O, NH, S$ (125)

$R = -(CH_2)_2-O-(CH_2)_2-O(CH_2)_2-$ (126)

complexes have been confirmed in a number of cases by X-ray diffraction studies. For instance, in the complex of (118, X = S), Cu_2L(imidazo-late)$(ClO_4)_3.H_2O$, the overall geometry of each copper is tetragonal; there are strong equatorial bonds from three nitrogens of the macrocycle and from one imidazolate nitrogen to each Cu(II) together with weaker perchlorate or water coordination in the respective axial positions (Drew, Cairns, Lavery & Nelson, 1980). The sulfur donors do not coordinate. The molecule is bent such that the planes incorporating each pyridine moiety intersect at 42°.

It is interesting that the 30-membered O_4N_6-donor macrocycle (126), on interaction with Cu(II), initially yields the mononuclear complex $[CuL](ClO_4)_2.H_2O$ in which the metal-ion is octahedrally coordinated by the six nitrogen atoms (Drew, McCann & Nelson, 1981). On treatment with excess free copper ion, the $[CuL]^{2+}$ species unfolds to yield binuclear $[Cu_2L](ClO_4)_4.2H_2O$. This complex has been used as a precursor for several derivatives containing small substrate ions (OH⁻, imidazolate, N_3^-, NCS⁻) intramolecularly bound between the copper centres. The structure of the OH⁻ bridged complex, $[Cu_2L(OH)(ClO_4)(H_2O)]$ $x[ClO_4]_2$, shows a Cu⋯Cu distance of 3.57 Å compared to 5.99 Å in the analogous imidazolate-bridged species. The ether oxygens of the macro-cycle do not coordinate but, in both structures, the metal ions are weakly bound to perchlorate or water oxygen atoms in axial positions.

For systems of the type just discussed, the inclusion of a bridging molecule or ion between the two metal centres is dependent on a number of factors. For example, the interaction of Pb(NCS)₂ with (126) leads to a binuclear product in which each Pb(II) is seven-coordinate – being bonded to three nitrogens and two oxygens from the ring, and two thiocyanate sulfurs above and below the ring (Drew, Rodgers, McCann & Nelson, 1978). Even though thiocyanate is a common bridging ligand in other metal systems, no such bridge forms in this complex. This undoubtedly reflects a balance of factors including the ability of each lead to attain seven coordination without bridge formation and the preference of Pb(II) for coordination to thiocyanate via sulfur (rather than via nitrogen).

Related behaviour occurs in the tetraazide binuclear copper derivative of (127, Y = O) for which an X-ray structure has been reported (Comar-mond *et al.*, 1982). This complex does not have a bridging group between the Cu(II) ions. Instead, the structure contains CuN_4 centres, with each Cu(II) having a square-pyramidal geometry formed by three ring nitro-gens and two terminally bound azide ions. One of the azide ions is directed outside the ring and the other lies parallel to the Cu–Cu axis but does not form a bridge between the copper atoms. The Cu–Cu distance is long at 5.973 Å and the copper ions are not magnetically coupled.

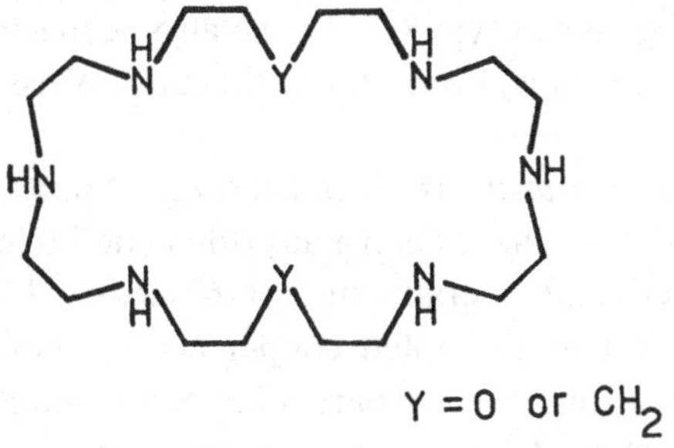

Y = O or CH$_2$

(127)

A feature of the metal-ion chemistry of these large ring macrocycles is thus the structural diversity which may occur from one system to the next. This diversity can result directly from small changes in the structure of the cyclic ligand and is also aided by the inherent flexibility of the large rings involved. It is clearly also influenced by the nature of the other ligands available for complex formation.

A solution equilibrium study. Potentiometric equilibrium studies have been used to investigate the stabilities of complexes of (127; Y = O) with Cu(II), Co(II), Ni(II) and Zn(II) (Motekaitis, Martell, Lecomte & Lehn, 1983). This ligand was shown to form stable 1:1 complexes in which between four and six of the nitrogens coordinate to the metal ion. The 2:1 (Cu:ligand) complex was demonstrated to be especially stable relative to those of the other ions. The lower stability of the binuclear complexes of Co(II), Ni(II) and Zn(II) is illustrated by the fact that precipitation of the respective metal hydroxides occurs in these cases as the pH of the solution is increased. In contrast, the binuclear Cu(II) species is stabilized as the pH is increased by the initial formation of a monohydroxo-bridged complex; at higher pH values a dihydroxo-bridged complex is formed. Similar investigations involving the dinuclear copper complexes of the

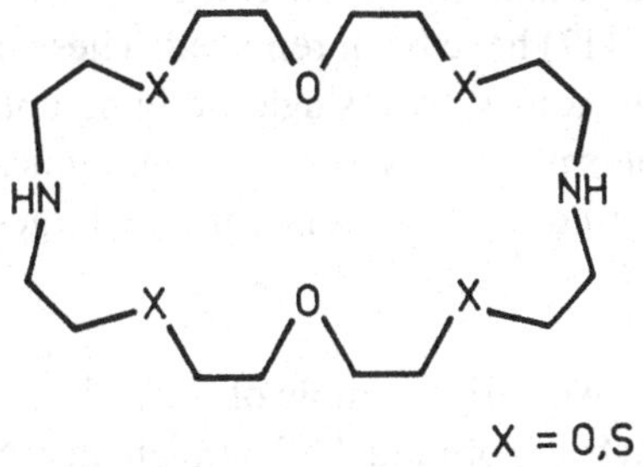

X = O,S

(128)

closely related ligands of type (128) have also been carried out (Comarmond *et al.*, 1982; Agnus, Louis, Gisselbrecht & Weiss, 1984).

The relation to copper proteins – a model study. Apart from the examples so far mentioned, there have been many other studies involving binuclear copper compounds and, overall, much of this activity has been stimulated by the observation that coupled copper centres have been shown to account for the properties of particular copper-containing biological molecules. Thus strongly antiferromagnetically coupled di-Cu(II) centres occur in met- and oxyhaemocyanin, laccase and related copper proteins although, for these systems, the nature of the bridging groups between the copper centres has not been firmly established.

In one study of this type, di-Cu(II) complexes of the aliphatic macrocycles (127) have been used in an attempt to model the properties of the natural systems (Coughlin *et al.*, 1984). For a number of these complexes, each Cu(II) is bound to an N_3-donor sequence of the ring while a bridging ligand occupies the fourth position in the respective equatorial planes of each copper. Other ligands occupy axial sites. The flexibility of the ring system permits the inclusion of bridges of different sizes. For an imidazolate-bridged species of this type, the Cu–Cu distance was found to be 5.86 Å (Coughlin *et al.*, 1979; Coughlin, Lippard, Martin & Bulkowski, 1980) whereas this is reduced to 3.64 Å when the bridge is a monohydroxo species (Coughlin & Lippard, 1981). Magnetic studies indicate that there is close agreement between the exchange coupling constants in this latter complex and those estimated to occur in laccase and oxyhaemocyanin. The synthetic species also exhibits an absorption maximum at 330 in the uv-visible spectrum which is characteristic of binuclear copper centres in biological molecules. These and other observations have been used to support the proposal that specific biological systems contain OR^- or OH^- bridges between their Cu(II) centres.

Two-ring systems incorporating a single bridge

A number of examples of the category of binucleating macrocycles illustrated by (117) have been reported. These derivatives contain two macrocyclic rings joined by a single bridging unit; this ligand type tends to yield complexes which show less communication between the metal centres than frequently occurs for the large-ring dinucleating species just discussed.

Some typical systems. An early example of a metal derivative of this class of ligand is (129) which was obtained (Cunningham & Sievers, 1973) by a

(129)   (130)

(131)

direct coupling reaction involving two molecules of the related monomeric macrocyclic derivative (130). The dimeric structure of the product was confirmed by an X-ray diffraction study. The ligand has not been isolated free of its metal ions in this case.

In another study it was demonstrated that a small yield of a related dimeric molecule (131), in which two cyclam subunits are linked through a covalent bond, is formed as a side product during the template synthesis of cyclam (Barefield, Chueng, Van Derveer & Wagner, 1981). A simple direct synthesis of the corresponding dimeric derivative of dioxocyclam (132) has also been reported (Fabbrizzi, Forlini, Perotti & Seghi, 1984) – see [3.3]. The interaction of Cu(II) with (132) has been investigated using potentiometric techniques. Adjustment of the ligand:metal ratio leads to the formation of both monometallic and dimetallic complexes. Insertion of a Cu(II) into a tetraaza subunit promotes simultaneous deprotonation of two amide groups. The uncharged di-Cu(II) species undergoes reversible oxidation to the corresponding di-Cu(III) species via two consecutive one-electron steps whose potentials differ by only 110 mV. Much redox chemistry of general interest, such as the oxidation or reduction of organic or bio-organic substrates, involves the transfer of a pair of electrons in a single step. Thus complexes such as (132), which contain

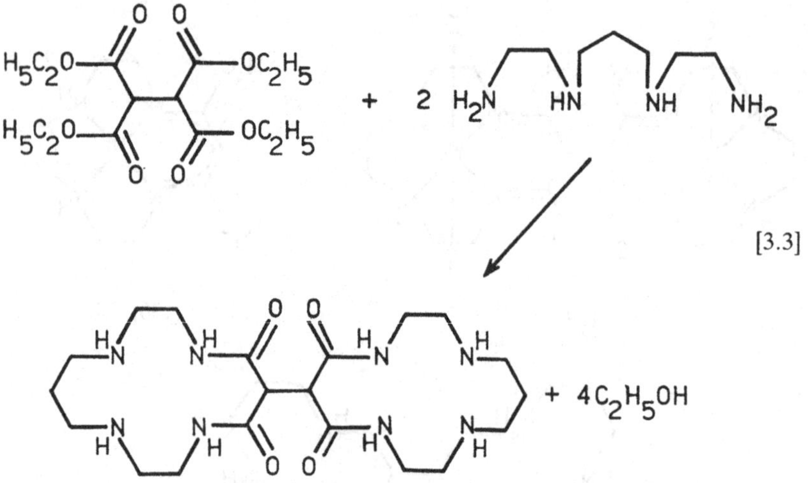

$$+ \quad 2 \; H_2N \quad HN \quad NH \quad NH_2$$

[3.3]

(132)

$$+ \; 4 C_2H_5OH$$

two connected metal centres exhibiting quasi-independent redox activity, may prove useful as two-electron redox reagents. It has been well documented (see Chapter 8) that the redox potentials of metal centres in cyclic systems can be tuned in a systematic manner by variation of structural elements associated with the ring; many one-electron cyclic systems, spanning a very wide potential range, have been identified. On connecting two such one-electron systems together as occurs in particular complexes of (132), the metal centres will be expected to act in parallel such that an effectively two-electron transfer will occur.

Reduction of (132) yields the analogous bis-cyclam derivative which, as expected, exhibits related chemical properties to its isomer (131).

Examples (129), (131) and (132) all contain one bond linking two macrocyclic rings. Ligands containing more elaborate linkages between the rings have also been synthesized. Typical of these is (133) (Fleisher *et al.*, 1973) which was obtained by the direct synthesis illustrated by [3.4]. Reaction of the free ligand with Cu(II) or Ni(II) acetate results in formation of the corresponding neutral dimetallic complexes. In contrast to this synthetic procedure, the related dimeric Ni(II) complexes of type (134, *n* = 2 or 3) may be obtained (Black, Vanderzalm & Wong, 1979) by *in situ* condensation of the appropriate organic precursors in the presence of nickel acetate in hot DMF.

Two further examples of this type are given by the di-Cu(II) complexes of (135) and (136). Ligand (135) forms a complex of composition $[Cu_2L](ClO_4)_4$ in which the Cu(II) ions were assumed to be contained in

(133)

[3.4]

(134)

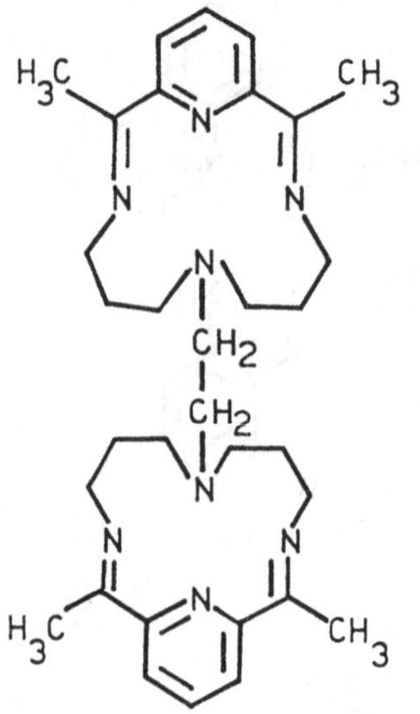

(135)

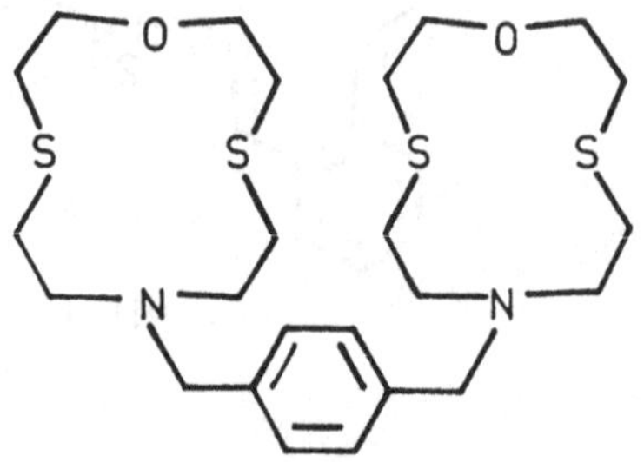

(136)

the respective macrocyclic rings (Murase, Hamade & Kida, 1981). Magnetic moment and esr measurements indicate that weak antiferromagnetic interaction occurs between the two copper centres. In view of this, it was postulated that the cyclic coordination units are arranged so as to face each other in the complex. By this means, direct interaction between the copper ions may occur. If the coordinated ligand remains approximately planar then the distance between the copper sites would appear to preclude intramolecular interaction via the dimethylene bridge. In accordance with this, ligand (136) yields a di-Cu(II) derivative in which the ligand maintains a non-folded configuration – the magnetic and esr properties of this complex indicate no interaction between the copper centres (Burk *et al.*, 1981). However, when this complex is dissolved in water at pH 6, it is transformed into a different species, which has been isolated. Magnetic studies indicate that this new product contains copper ions which are strongly antiferromagnetically coupled. An X-ray structure shows that the ligand has now assumed a folded 'ear muff' configuration which is similar to that proposed for the complex of (135). However, in the complex of (136), a hydroxo bridge is formed between the copper ions. Each Cu(II) achieves an approximately square-pyramidal geometry with the basal plane defined by the ONS$_2$-set of one ring and the axial position occupied by the OH$^-$ bridge.

A linked porphyrin system. Urea-linked porphyrin rings have also been synthesized (Landrum, Grimmett *et al.*, 1981). For example, imidazolate-

(137)

bridge species of type (137) have been prepared containing a variety of homo and hetero metal-ion pairs. The Mn(II)/Co(II) species was proposed to be a potential spin model for cytochrome oxidase even though the natural system is an Fe(III)/Cu(II) species. Nevertheless, in both systems there is an $S = \frac{5}{2}$ metalloporphyrin interacting with an $S = \frac{1}{2}$ metalloporphyrin. From the measured small value of the coupling constant ($J = 5$ cm^{-1}) for the antiferromagnetic coupling between the metals in the model system, it was concluded to be unlikely that histidine acts as a bridging ligand between the metal-ion pair in cytochrome oxidase.

'Co-facial' dimers

The final category of binucleating ligands are those which incorporate more than one linkage between the cyclic subunits [see (118)]. The additional linkages between rings will tend to reduce the flexibility of the system and, for example, usually enable the relative positions of the metal centres to be controlled more precisely. The most common examples in this category have been termed 'co-facial' dimers.

Co-facial dimers consist of two simple macrocyclic systems connected by linkages such that the respective cyclic portions are constrained to 'stack' one above the other. Such molecules define three cavities: two within the macrocyclic subunits and a third between these subunits.

A limited number of such ligands have been reported; several of these are crown ether derivatives and will be discussed in Chapter 4.

Some representative co-facial systems. The series (138)–(140) of mixed thiaaza donor ligands have been synthesized and all three have been shown to form binuclear complexes with transition metal ions (Lehn, 1980). In these, each metal ion is associated with one cyclic subunit as shown in (141) for (138). Complexes containing two Cu(II) ions or two Cu(I) ions have been obtained with each of (138)–(140) while (139), the

(138)   (139)   (140)

(141)

unsymmetrical system, also gives a mixed Cu(II)/Cu(I) complex. The structure of the di-Cu(II) complex of (138) has been determined (Louis, Agnus & Weiss, 1978). Each copper is bound to an ON_2S_2-donor set in a distorted tetragonal prismatic arrangement such that both ions lie approximately 0.34 Å out of the respective N_2S_2-planes (towards the respective axial oxygen atoms). The displacement of the metals towards the central cavity results in this system showing structural similarities to the 'cage' complexes discussed later in this chapter.

The synthesis of a further group of co-facial dimers containing all-nitrogen donor atoms have been described (Busch, 1980). For example, Ni(II) species of type (142 with R = $-(CH_2)_2-$, $-(CH_2)_2$, $-(CH_2)_4$, *m*-xylene, or C_7-fluorene) have been isolated (Busch *et al.*, 1981; Herron *et al.*, 1983); these nickel dimers contain the metal ions in square-planar environments.

(142)

A number of binuclear iron complexes have also been isolated (with a neutral base attached to each metal in an axial position). The iron complexes undergo net two-electron redox reactions with dioxygen to yield products containing two identical low-spin Fe(II) metal sites; superoxide or peroxide are simultaneously generated. Remarkably, the reaction can be partially reversed by removal of O_2 from the system by, for example, flushing with N_2 in a mixed aqueous solvent at 0 °C.

Co-facial porphyrin dimers. A number of dimetallic co-facial porphyrin systems involving a variety of bis linkages between the porphyrin rings have been studied for some time (Chang, 1977; Collman, Elliot, Halbert & Tovrog, 1977; Kagan, Mauzerall & Merrifield, 1977). For example, in work by Chang and coworkers, addition of dioxygen to a co-facial haem dimer led to virtual instantaneous oxidation of the haems at −45 °C. The rate of oxidation is at least 1000 times greater than for (monomeric) myoglobin. In the monomeric systems, the rate-limiting step is the formation of the peroxo-bridged complex; for the co-facial dimer the rate is much increased because of the favourable relative positions of the Fe(II) ions for bridge formation. That is, reduction of dioxygen is enhanced because it is sandwiched between two Fe(II) ions and thus is in an ideal situation to accept an electron from each Fe(II) to form the O_2^{2-} bridge.

In a subsequent study of this type (Durand, Bencosme, Collman & Anson, 1983), dimers of type (143) were investigated as potential redox catalysts for the four-electron reduction of dioxygen to water (via peroxide). The Co(II)/Co(II) dimer is an effective catalyst for the elec-trochemical reduction of dioxygen: once again the dioxygen binds

(143)

between the pair of porphyrin rings such that interaction with both Co(II) ions occurs. However, for a number of related mixed-metal dimers containing Co(II) as one of the metals, the second metal was found to be catalytically inert. In these cases, the redox behaviour did not differ greatly from that observed for the corresponding monomeric Co(II) porphyrin.

3.5   Bicyclic-ligand complexes incorporating a void adjacent to the metal ion

Starting with Ni(II) complexes of type (144), Busch and coworkers have prepared a large number of bicyclic species of type (145) which were named 'lacunar' complexes (Korybut-Daszkiewicz *et al.*, 1984). Equation [3.5] outlines a typical synthetic procedure although variations of this procedure have also been employed. In these complexes, R^2 is usually H or CH_3, R^3 can be one of a large number of substituents while R^1 is either *m*-xylene or an aliphatic bridge of type $-(CH_2)_n-$ (where $n = 4$–7). The bicyclic ligands may be removed from Ni(II) by treatment of the respective complexes with HBr in methanol and then transferred intact to such ions as Co(II) and Fe(II). The latter complexes display exceptional activity as reversible carriers for dioxygen in which the O_2 binds to the metal and is contained in the void of the ligand. Other related oxygen-carrying complexes of this type which mimic the behaviour of natural systems will be discussed in Chapter 9.

The synthetic procedures developed for the present bicyclic systems have enabled steric control of the size of the available void to be achieved and, for example for the smallest voids, the ability of the complexes to reversibly bind O_2 is lost (Herron, Chavan & Busch, 1984). Similarly, by variation of the ligand substituents, some control of the electronic environment within the void can also be effected. Overall, this ability to 'tune' the cavity both sterically and electronically very greatly enhances the versatility of this group of ligands for acting as hosts for small molecules such as O_2 or CO.

In other studies, a group of related ligands incorporating considerably larger cavities have been synthesized (Takeuchi, Busch & Alcock, 1981). These 'vaulted' systems were isolated as their diamagnetic Ni(II) complexes and are of type (146). Using ^{13}C nmr, it has been demonstrated that these nickel complexes act as hosts for the formation of inclusion complexes in aqueous solution (Takeuchi & Busch, 1983). A variety of different alcohols and phenols enter the hydrophobic cavities of these systems. After inclusion, the signs and magnitudes of the nmr chemical shift changes have been accounted for by a model in which the guest molecule replaces water from within the cavity. The binding of these guests in the cavity is relatively weak with the binding constants (K) for the series of alcohols all being less than 10 mol^{-1} dm^3. The guest-host association is believed to occur largely through hydrophobic interactions.

Complementary ^{1}H nmr relaxation studies involving an analogous

(146)   etc.

vaulted Cu(II) complex have been performed (Kwik, Herron, Takeuchi & Busch, 1983). The studies demonstrate that the organic guests bind in preferred orientations – the hydrophobic portion of each guest is found to lie inside the hydrophobic cavity close to the metal while the hydrophilic (hydroxy) portion extends into the bulk aqueous solution.

3.6   The cage macrocycles

An important development in macrocyclic chemistry, dating from 1968, was the synthesis of the first of a series of three-dimensional 'cage' ligands which are able to completely encapsulate a coordinated metal ion. The potential of this macropolycyclic type to provide a sterically-controlled coordination cavity such that complexes with unusual properties might result was recognized early and the development and study of new cage (or cryptand) systems has proceeded since this time. In particular, cages might be expected to exhibit both enhanced specificity and stability for individual metal ions and the general structural, spectroscopic, and electrochemical features of such systems are all of considerable interest. Further, in theory, cage chemistry provides the prospect of 'fixing' a range of metal ions in similar coordination environments and thus facilitating a comparative study of the metal-ion chemistry under similar conditions. This is usually less possible with ligand systems which are not as sterically defined as the cages.

Unsaturated cage systems

Cages derived from oxime precursors. In the initial synthesis of a cage (Boston & Rose, 1968), tris(dimethylglyoximato)cobalt(III) [where dimethylglyoximato (dmgH) = (147)] was reacted with boron trifluoride

(147)

(148)

to produce the Co(III) complex of (148). In the formation of the cage, two BF_3 molecules react along the C_3-axis of the octahedral tris-dmgH complex to 'cap' each set of three oxime functions.

Apart from the boron halides, a number of other Lewis acids such as $SnCl_4$, $SiCl_4$, and H_3BO_3 can also be used as capping reagents in reactions of this type (Boston & Rose, 1973). A range of related complexes of Co(III), Co(II) and Fe(II) have been reported (Jackels & Rose, 1973). In an alternative synthetic procedure, the metal salt, dmgH₂, and BF_3 or $B(OH)_3$ are reacted together in an alcohol. In this manner, metal cages capped by BX groups (where X = F, HO, CH_3O, C_2H_5O, i-C_3H_7O, or n-C_4H_9O) have been isolated: the alkoxy groups are derived from the solvent used. All of these α-diimine products are quite stable chemically.

The structures of the parent (148; X = F) Co(III) and Fe(II) complexes have been determined by X-ray diffraction. Both compounds have coordination geometries which are intermediate between a trigonal prism and an octahedron. For example, the Co(III) species shows a twist angle (ϕ) between the two triangular planes containing the donor atoms of 45° relative to the trigonal prismatic arrangement (for which $\phi = 0°$) (Zakrzewski, Ghilardi & Lingafelter, 1971); a value of $\phi = 60°$ would indicate a fully octahedral complex. It is of interest that the structure of the Co(II) analogue gives a ϕ value of 8.5°. The tendency to form complexes which approach a trigonal prismatic geometry has been a special feature of several cage systems and has provided an avenue for more fully exploring the electronic and other properties associated with this little-observed coordination geometry.

A related procedure to that discussed for the tris-dmgH system has been employed to cap one end of Fe(II), Co(II), Ni(II) and Zn(II) complexes of the tripod ligand, tris(2-aldoxime-6-pyridyl)phosphine

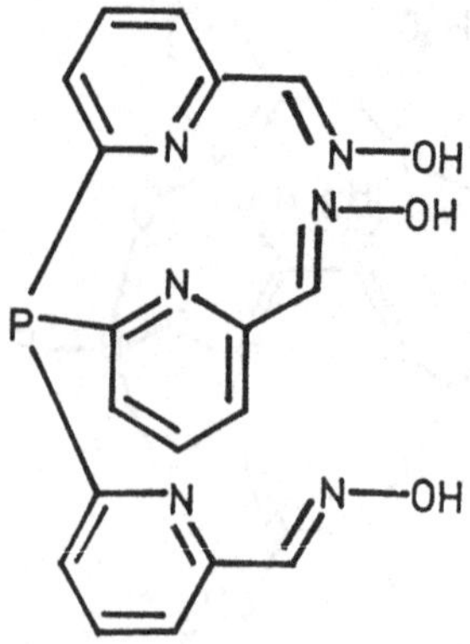

(149)

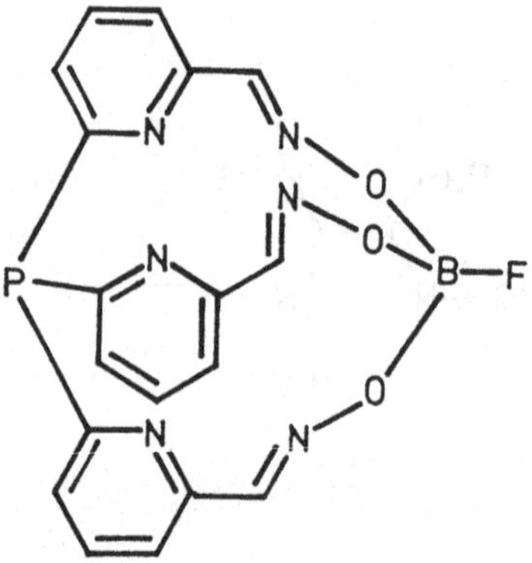

(150)

(149). The products are the corresponding complexes of the cage, fluoroborotris-(2-aldoximo-6-pyridyl)phosphine (150) (Parks, Wagner & Holm, 1971). Both boron trifluoride and the tetrafluoroborate ion have been used successfully as capping reagents for this system. The metal complexes are all six-coordinate; this ligand framework appears quite successful in imposing trigonal prismatic or near trigonal prismatic stereochemistries on the respective encapsulated metal ions (Larsen *et al.*, 1972; Churchill & Reis, 1973).

Two further unsaturated cages. Each of the systems discussed so far involves reaction of an electrophilic reagent with non-coordinated nucleophiles appended to metal-bound ligands. In contrast, in the following synthesis, cage formation occurs via an internal rearrangement of an Fe(II) complex of type (151) (Herron *et al.*, 1982). Complexes of type (151) have already been discussed in Section 3.5. Treatment of these

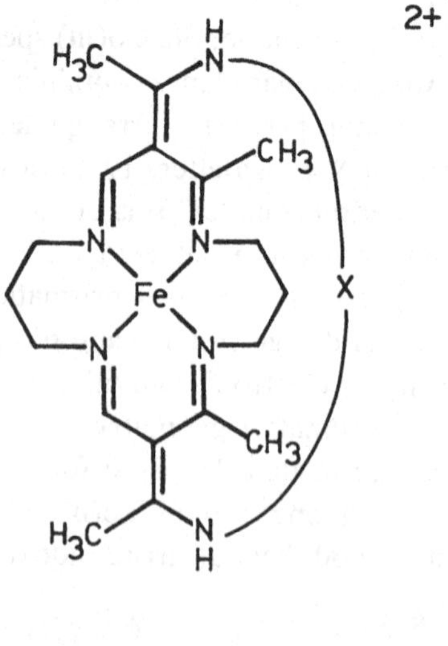

(151)

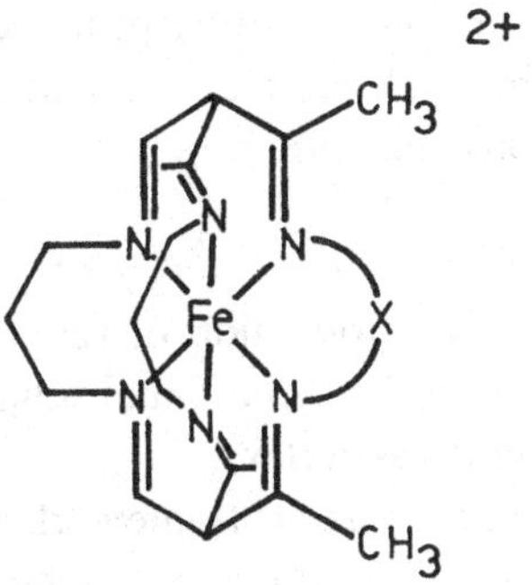

(152)

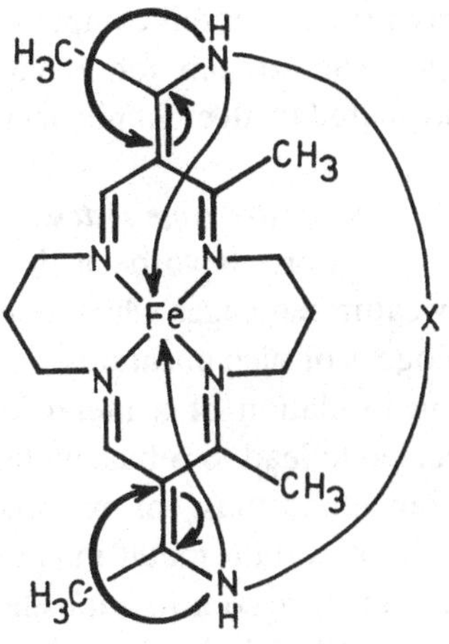

(153)

complexes with base in hot methanol results in a tautomeric rearrangement of the conjugated π-systems of the macrocycle followed by coordination of the previously uncoordinated amine groups to yield (152). During the course of the rearrangement, transfer of the amine protons to the γ-carbon atoms of the ring occurs – see (153). The factors influencing this rearrangement are not well understood. The X-ray structure of one Fe(III) complex of this type has been determined. The six amine groups form a slightly distorted octahedron with the trigonal twist being only five degrees in this case.

The interesting N_6-case (154), incorporating three linked 1,10-phenanthroline moieties, has also been prepared as its sodium complex. An

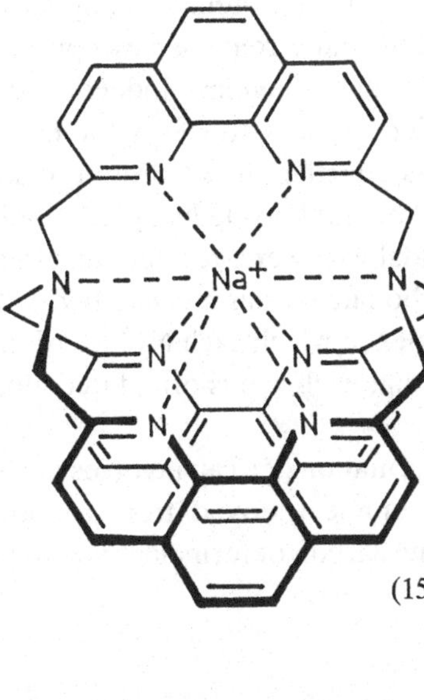

(154)

X-ray structure of this compound shows that the ligand adopts a propeller shape. The sodium ion is contained in the molecular cavity and is coordinated to all eight nitrogen atoms (Caron *et al.*, 1985).

Saturated cage systems

Efforts have been directed towards the production of related fully saturated cages which may be considered to be three-dimensional analogues of such unsaturated monocyclic ligands as cyclam.

Encapsulation of a metal ion by a *saturated* organic framework is expected to lead to robust metal derivatives which are stable over a wide pH range and thus, for example, inhibit the hydrolysis which is characteristic of certain metal ions in aqueous solution. In this manner, the non-hydrolytic coordination chemistry of these ions in solution becomes accessible. Similarly, the redox chemistry of such encapsulated ions is of special interest, since there exists the prospect that the saturated organic shell might 'insulate' the metal ion to a greater or lesser degree from the surrounding medium and hence markedly influence electron transfer reactions.

Saturated N-donor cages. The synthesis of the first N_6-donor saturated cage arose from observations concerning the extraordinary stabilities towards hydrolysis of a number of imine ligands coordinated to Co(III) (Sargeson, 1979). Such complexes can be isolated from concentrated hydrochloric acid even though the parent imines are completely hydrolysed under these conditions. This stability towards hydrolysis of the imine when bound to Co(III) presumably reflects both the kinetic inertness of the Co–N bond and the absence of an unbound lone pair of electrons on the imine nitrogens at which protonation might occur. Nevertheless, the carbon of the bound imine remains susceptible to attack by nucleophiles such as CN⁻ and carbanions. Indeed, the Co(III) appears to activate the bound imine towards attack by the nucleophiles just mentioned (but is much less effective than a proton in activating it towards attack by water). Based on this knowledge, $[Co(1,2\text{-diaminoethane})_3]^{3+}$, ammonia, and formaldehyde were reacted together in water in the presence of lithium carbonate (Creaser *et al.*, 1977). The product of this reaction is a cage complex in which each N_3-face of the starting complex has been capped such that each cap is orientated along the C_3-axis [see (155)].

The first step in the formation of a cap involves nucleophilic attack of a deprotonated primary amine (derived from a coordinated ethylenediamine) on the carbonyl carbon of formaldehyde to yield a bound imine

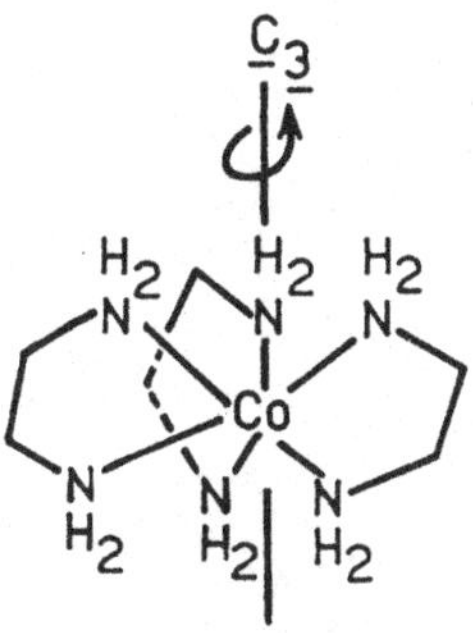

(155)

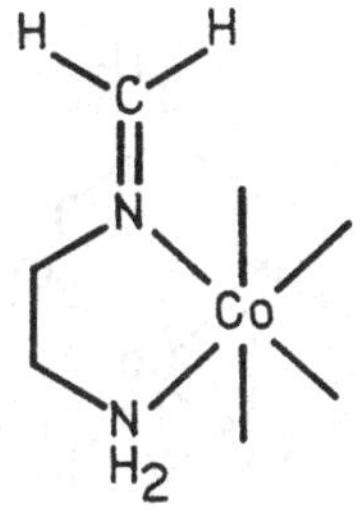

(156)

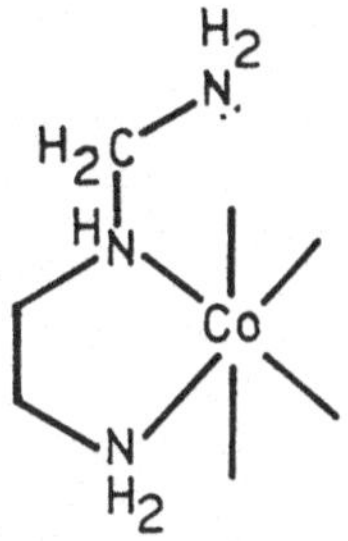

(157)

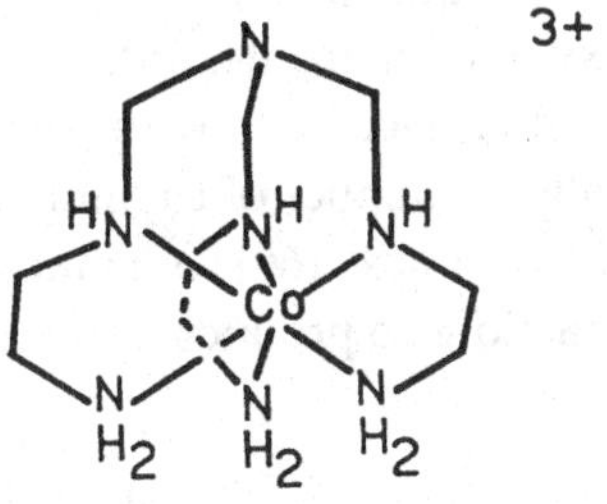

(158)

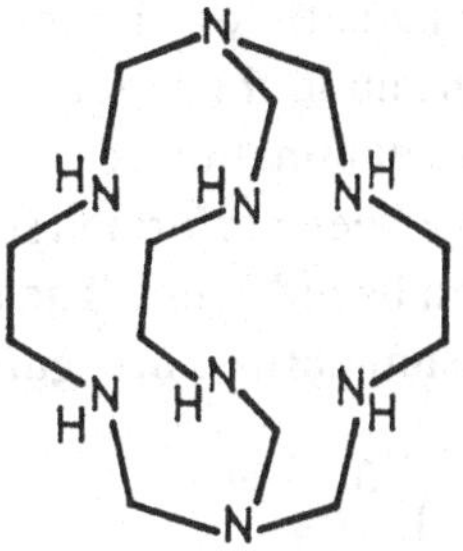

(159)

(156). The imine carbon is then attacked by an ammonia to yield a *gem* diamine (157). Further attack by the lone pair of the *gem* diamine on the carbon of a second coordinated imine leads to formation of the second link of the bridgehead. Finally, the process is repeated using the (now) secondary amine to produce the trigonal cap (158). Repetition of the entire process at the remaining N_3-face results in complete encapsulation of the metal-ion to yield (159); this cage was given the name 'sepulchrate' in keeping with the necrotic cryptate nomenclature (Creaser *et al.*, 1982).

X-ray diffraction studies of the Co(II) and Co(III) sepulchrate complexes indicate that they have similar structures although, as expected, the M–N bond lengths are shorter (1.99 Å) in the Co(III) complex than in the Co(II) species (2.16 Å). The Co(II) complex was obtained by direct reduction of the Co(III) analogue using zinc dust. A striking feature of the syntheses of these species is that the reactions proceed with retention of the configuration of the $[Co(1,2\text{-diaminoethane})_3]^{3+}$ starting complex.

The synthetic procedure was subsequently extended by substitution of a number of other molecules for ammonia in the capping procedure

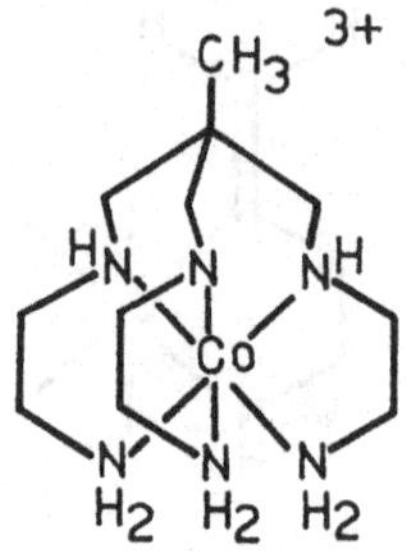

(160)

(Geue *et al.*, 1984). Thus, use of nitromethane leads to a nitromethylated-capped cage. Reduction of the nitro group to an amine may be achieved using Zn/H$^+$, and nitrosation of the amine in aqueous solution provides an intermediate for obtaining the hydroxy and chloro derivatives. The chloro group in the latter compound can be replaced by hydrogen via a reductive elimination reaction. Additionally, a number of capping reactions have been performed starting from the complex (160) containing the trifucated ligand, 'sen'. The use of such reactions to produce a variety of cages is summarized in Figure 3.1.

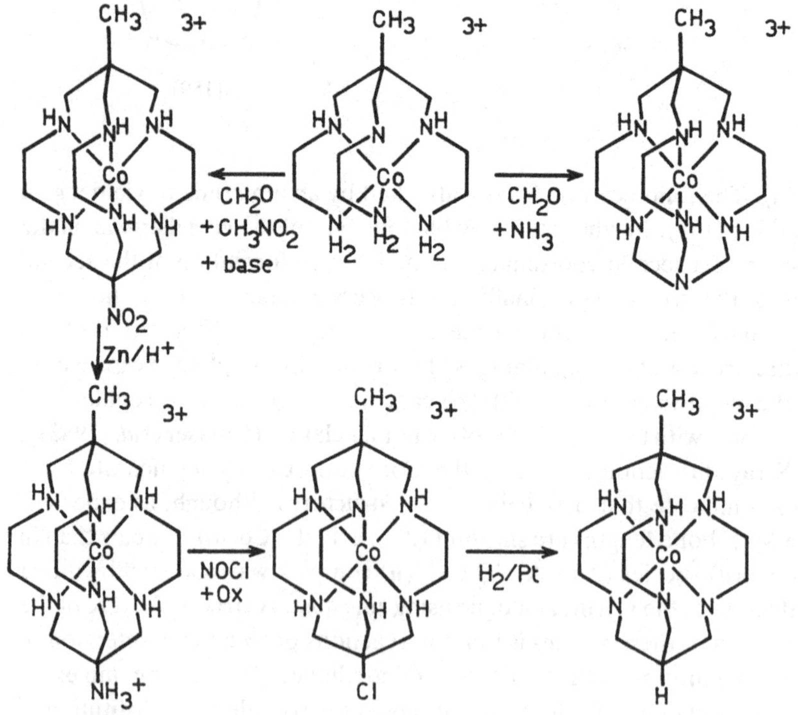

Figure 3.1. Synthetic pathways for obtaining unsymmetrically-capped cages.

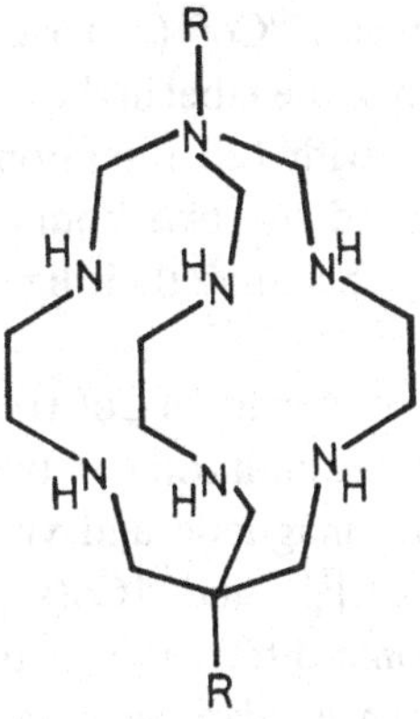

R = H ('sarcophagine')

= NH$_2$

= NO$_2$

(161)

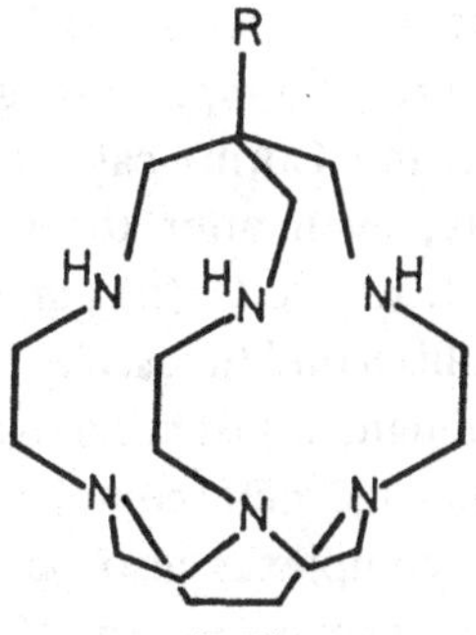

(162)

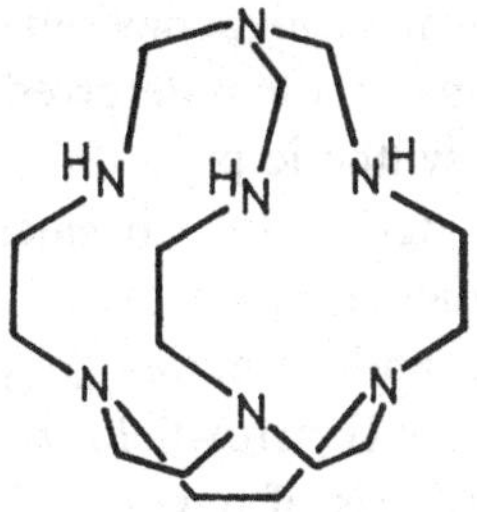

(163)

Other N$_6$-cages also prepared include symmetrical derivatives of type (161), the unsymmetrically capped species (162) and (163) (Boucher *et al.*, 1983; Bond, Lawrance, Lay & Sargeson, 1983; Hammershøi & Sargeson, 1983) and a number of related cages derived from 1,2-diaminocyclohexane rather than 1,2-diaminoethane (Geue, McCarthy & Sargeson, 1984).

The overall high symmetry of the cages results in their nmr spectra being, usually, quite simple – for example, [Co(sepulchrate)]$^{3+}$ gives just two ^{13}C-nmr resonances (of equal intensity) corresponding to the two carbon atom types in the molecule. The spectral simplicity is also a reflection of the stereo-specificity that is a feature of cage formation for these systems. In the cage complex, each of the bound nitrogen centres is chiral and the metal centre is also chiral. Hence a considerable number of isomers is theoretically possible. In fact, starting from either chiral form of [Co(1,2-diaminoethane)$_3$]$^{3+}$, there are 16 isomers which could occur; however only one is observed! This remarkable result is a direct consequence of the sterically-required stereospecificity of each step of the capping reaction. Similarly, once formed, the cage is a tight fit about the cobalt ion and the resultant inflexibility inhibits the formation of different conformers in solution. Indeed, reduction of [Co(sepulchrate)]$^{3+}$ to the analogous Co(II) complex followed by treatment with $^{60}Co^{2+}$(aq) over 17 hours at room temperature gave no incorporation of the label in the cage. In fact, the Co(III) cage could be regenerated with full retention of chirality, even after the Co(II) species had stood for two hours. In contrast, simple Co(II) amine complexes usually exchange their ligands on a millisecond timescale.

Although, as just pointed out, a number of the properties of Co(II) cage and non-cage metal complexes may be very different, similarities between related complexes also occur. For example, the magnetic and visible spectral properties of $[Co(1,2\text{-diaminoethane})_3]^{3+}$ and $[Co(sepulchrate)]^{2+}$ are essentially the same, although some differences in their circular dichroism spectra are apparent. Differences also occur in the electrochemical behaviour of cage and non-cage species but further mention of these is deferred until Chapter 8.

The synthetic procedures developed have been used to obtain other metal derivatives besides cobalt. Thus, for example, $[M(1,2\text{-diaminoethane})_3]^{n+}$ [M = Pt(IV), Rh(III), or Ir(III)] react with formaldehyde and ammonia to yield the corresponding (inert) sepulchrate complexes (Harrowfield *et al.*, 1983; Boucher *et al.*, 1983). A major aid to the synthesis of new metal derivatives has been the development of a procedure for demetallating cage complexes. The metal-free cages are generated by prolonged heating of the corresponding metal derivatives in strong acid at elevated temperature. The availability of the free cages provided a means for the direct synthesis of complexes of other metal ions. Using both direct and *in situ* procedures (sometimes also coupled with redox reactions), cage derivatives of each of the following metal ions have been prepared: Mg(II), Ga(III), In(III), Cr(III), V(III), V(IV), Mn(II), Mn(III), Fe(II), Fe(III), Co(II), Co(III), Ni(II), Ni(III), Cu(II), Zn(II), Cd(II), Hg(II), Ag(II), Rh(III), Ir(III) and Pt(IV) (Sargeson, 1984). The X-ray structures of many of these complexes have been determined: once again, the observed coordination geometries span the range from near octahedral to almost trigonal prismatic.

Mixed-donor cages. The studies just discussed have been extended to include sulfur-containing cage derivatives. Figure 3.2 illustrates the synthetic procedure used to produce new cages of this type (Gahan, Hambley, Sargeson & Snow, 1982). In general, the cobalt complexes of these cages show many similar properties (such as resistance to racemization) to those of the hexaaza analogues.

In other studies, Lehn and coworkers have produced a range of mixed-donor 'cryptands' using direct synthetic techniques (usually at high

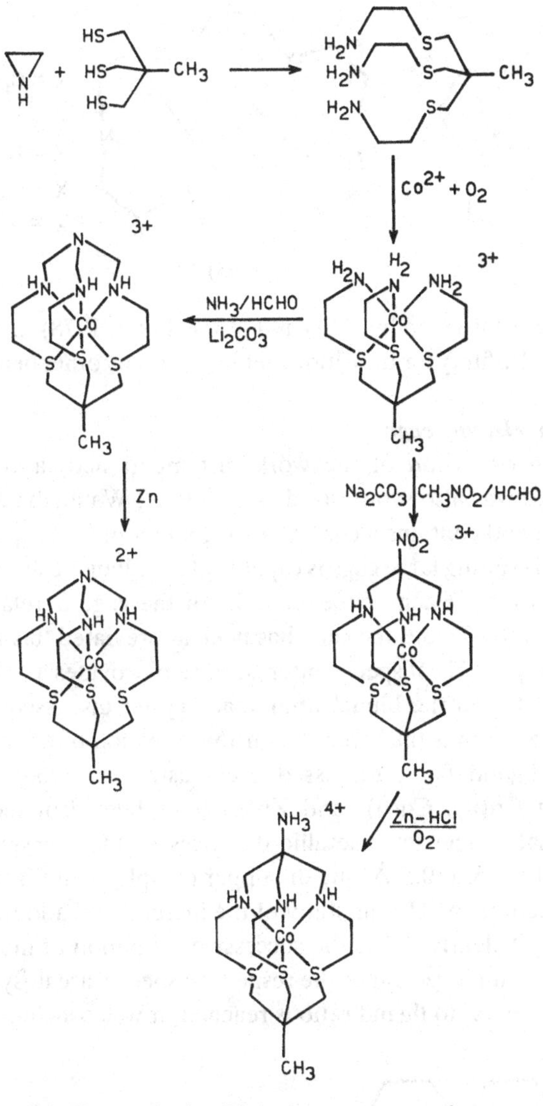

Figure 3.2 The synthesis of N_3S_3-donor cage complexes.

dilution). Although the overall yields have generally been lower than for the systems prepared by Sargeson and coworkers, the direct procedure has enabled cryptands of considerable structural diversity to be isolated (Lehn, 1978). The majority of these ligands contain donor sets in which polyether donor functions predominate and these will be discussed in the next chapter. However, ligands of this general type, incorporating higher numbers of nitrogen and/or sulfur donors, have also been obtained; a

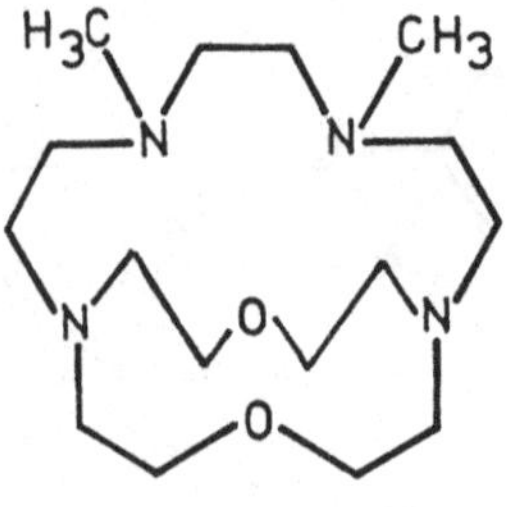

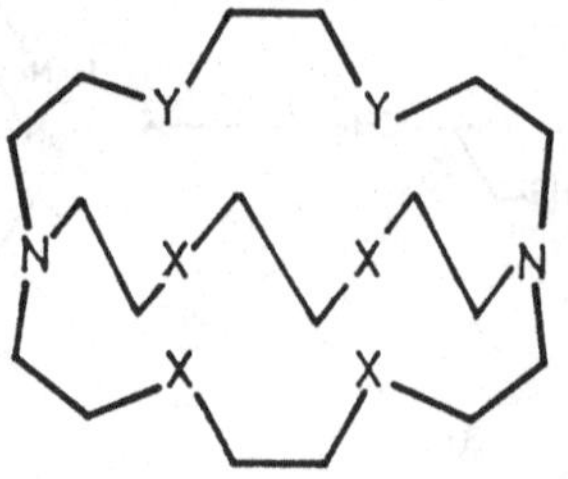

$$X = O, \quad Y = NCH_3$$
$$X = NCH_3, \quad Y = O$$
$$X = O, \quad Y = S$$
$$X = S, \quad Y = O$$
$$X = Y = S$$

(164)   (165)

selection of these is given by (164) and (165) (Lehn, 1978). Such ligands show increased affinity for transition metals and many other heavy metals.

A binucleating cage

In an extension of the work just mentioned, a cryptand of structure (166) was also synthesized (Lehn, Pine, Watanabe & Willard, 1977). This ligand contains a coaxial arrangement of two tripod N_4-subunits and the resulting large cage is capable of simultaneously encapsulating two metal ions. Each of the subunits of the cage is related to the N_4-ligand 'tren' (167) and the cage has been abbreviated 'bis-tren'. This category of cryptand is of special interest since it provides a link between the cage ligands and the binucleating macrocycles discussed in Section 3.4. Indeed, bis-tren is the three-dimensional version of the monocyclic binucleating ligand (127) discussed previously. For (166), binuclear complexes of Co(II), Cu(II), and Zn(II) have been isolated as their perchlorate salts. The intermetallic distances in these cryptates were estimated to be 4.5 ± 0.5 Å; the di-copper complex exhibits only weak Cu–Cu interaction. A ^1H nmr study of the incremental addition of ZnI_2 to (166) in D_2O clearly shows the successive formation of mononuclear and binuclear complexes. Since the respective spectra are unsymmetrical before the 2:1 metal to ligand ratio is reached, it was concluded that the

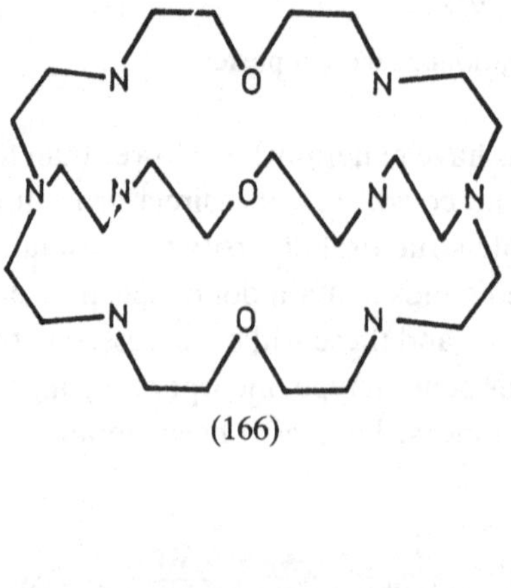

(166)

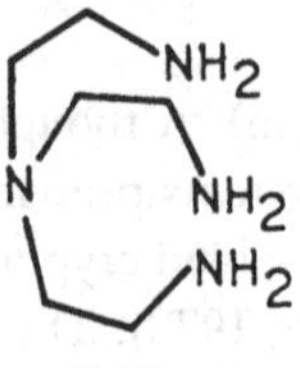

(167)

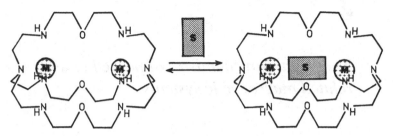

Figure 3.3. Schematic illustration of the manner by which a small molecule or ion may bridge the metal centres.

side to side exchange of the zinc ion within the cage is slow with respect to the nmr time scale. The data also indicate that the exchange between free and complexed Zn(II) is slow. There is strong evidence that small molecules or ions such as solvent, cyanide or azide, may also be incorporated in the cage to form bridges between the metal-ion sites. Such species have been termed 'cascade' complexes by Lehn; aspects of their formation are illustrated in Figure 3.3.

The solution chemistry of the Co(II), Ni(II), Cu(II), and Zn(II) species formed by bis-tren has been investigated by potentiometric (pH) equilibrium measurements (Motekaitis, Martell, Lehn & Watanabe, 1982). The results clearly demonstrate the formation of 1:1 and 1:2 complexes in each case and, as expected from consideration of electrostatic effects, the equilibrium constant for the binding of the second cation within the cage is less than for the first. Series of protonated species as well as hydroxo-metal species for both the mononuclear and binuclear complexes are also formed. The hydroxo groups bridge the bound metal ions in the binuclear complexes to form cascade complexes and, indeed, the affinities for hydroxide ion of the binuclear bis-tren complexes are very much higher (10^2–10^6) than occur for the mononuclear cryptates. The binuclear Cu(II) and Co(II) cryptates also combine with a second hydroxo ligand – it appears that the cryptate structure helps stabilize double bridge formation in these $LM_2(OH)_2$ species.

The binuclear Co(II) cryptate (monohydroxo bridged) was also found to combine reversibly with dioxygen yielding a doubly-bridged cascade complex containing both hydroxo and peroxo bridges within the cryptate structure.

Besides (166), other binucleating cryptands have been synthesized (Lehn, 1980). The availability of systems such as these once again provides a means for fixing a pair of metal ions in close proximity in a well-defined environment. Such species are ideal for the types of studies involving binuclear systems outlined in Section 3.4.

4

The metal-ion chemistry of polyether and related macrocyclic systems

4.1 The crown polyethers

Some preliminaries

The macrocycle types discussed so far tend to form very stable complexes with transition metal ions and, as mentioned previously, have properties which often resemble those of the naturally occurring porphyrins and corrins. The complexation behaviour of these macrocycles contrasts in a number of ways with that of the second major category of cyclic ligands – the 'crown' polyethers.

In 1967, a seminal paper describing the syntheses of thirty-three cyclic polyethers, of which (168)–(172) are typical, was published (Pedersen, 1967). All are simple monocyclic rings – a general macrocyclic category sometimes termed the coronands (Weber & Vogtle, 1980). In Pedersen's study, the first ligand to be prepared was (170) and because of the appearance of its molecular model and its ability, on coordination, to 'crown' a metal ion, this and other members of the series were referred to as crown compounds. The trivial names consist of, in order: (i) the number and kind of attached hydrocarbon rings, (ii) the number of atoms in the polyether ring, (iii) the class name, crown, and (iv) the number of oxygens in the polyether ring. Thus ligand (170) is referred to as dibenzo-18-crown-6.

Following the original paper, reports of the synthesis of new crowns and crown-like molecules proliferated. A typical property of these systems is their ability to form stable complexes with the alkali metal and alkaline earth ions. Prior to the synthesis of the crowns, the coordination chemistry of the above ions with organic ligands had received very little attention. A further impetus to the study of such complexes was the recognition of the important role of Na^+, K^+, Mg^{2+} and Ca^{2+} ions in biological systems.

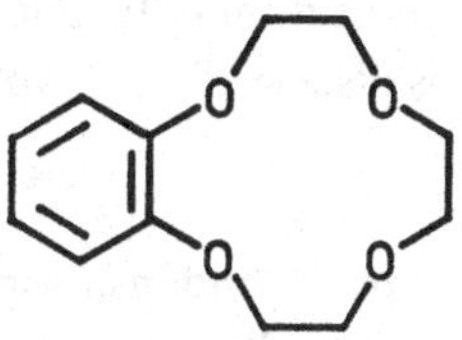

Benzo-12-crown-4

(168)

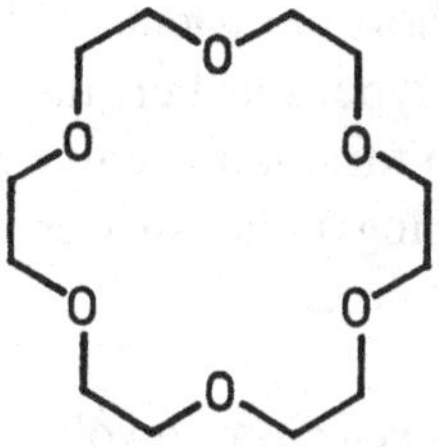

18-crown-6

(169)

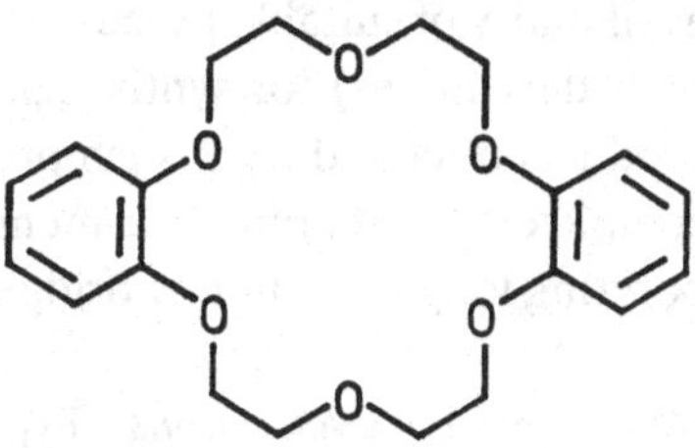

Dibenzo-18-crown-6

(170)

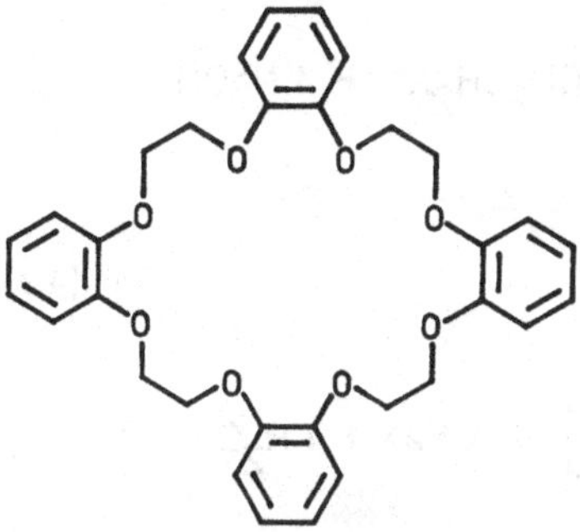

Tetrabenzo-24-crown-8

(171)

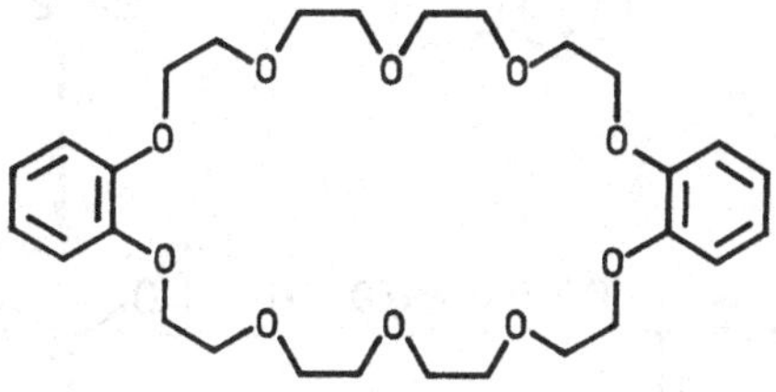

Dibenzo-30-crown-10

(172)

Apart from metal ions, many of these polyether compounds also exhibit complexing ability (and often specificity), for a range of other inorganic and organic cations as well as for a variety of neutral molecules. Complexes such as these, which contain species incorporated in the macrocyclic cavity, are usually known as **inclusion complexes**; the general area covering the binding of all types of substrates in molecular cavities often being referred to as **host-guest** chemistry. The present chapter deals with the synthesis and metal-ion binding of crowns and related systems.

The formation of complexes of non-metallic guests by crowns as well as by other types of cyclic hosts is discussed in Chapter 5.

Crown ether synthesis

General considerations. Condensation reactions, often at medium to high dilution, have usually been used to obtain new crown polyethers (Hiraoka, 1982; Gokel & Korzeniowski, 1982). For the majority of such crowns, two carbon atoms link consecutive ether oxygens in the respective rings. This structural feature is, in part, a reflection of the ready availability of suitable precursors (such as catechol or ethylene oxide and their derivatives) for synthesizing rings of this type. Further, the cyclic products obtained from such precursors tend to form more stable metal complexes (containing five-membered chelate rings) than species incorporating longer or shorter bridges between the oxygens.

Representative cyclizations. Typically, a given cyclization involves nucleophilic displacement of a halide or tosylate by alkoxide or phenoxide ion (that is, a Williamson synthesis) even though such reactions are frequently quite slow. Schemes [4.1], [4.2] and [4.3] give representative

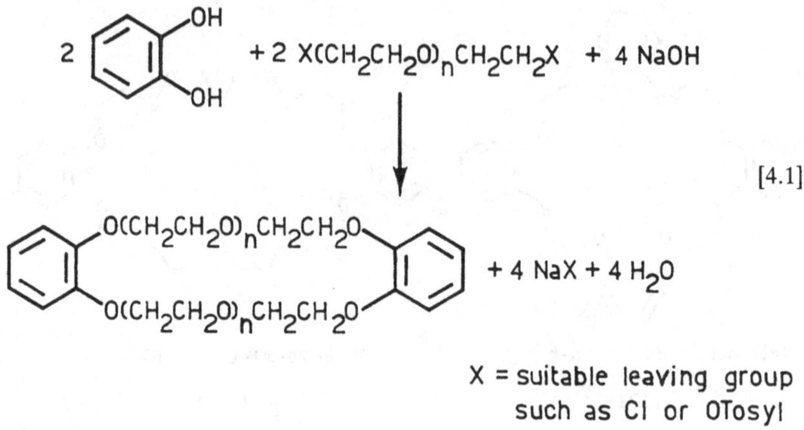

[4.1]

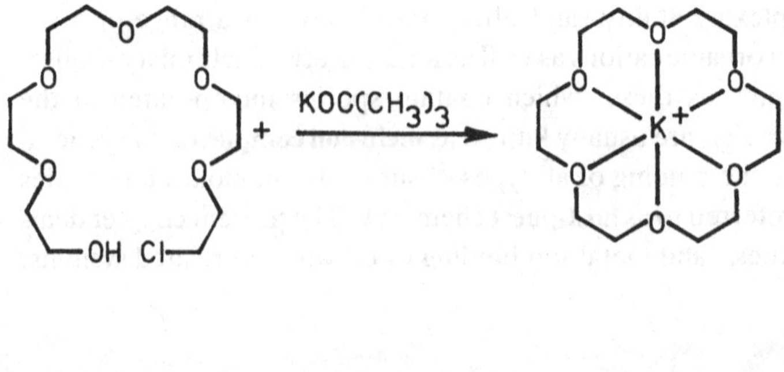

X = suitable leaving group such as Cl or OTosyl

[4.2]

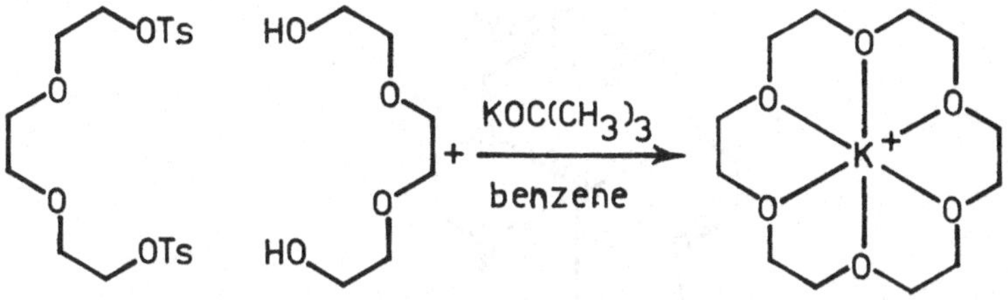

[4.3]

procedures. Products of the type shown in [4.1] may be converted into the corresponding cyclohexane derivatives by hydrogenation, typically in butanol, over a ruthenium catalyst.

A range of crowns incorporating furan or tetrahydrofuran heterocyclic rings have been synthesized. An example is given by (173). The general preparation of rings of this type has involved condensation of 2,5-bis(hydroxymethyl)furan with a polyether ditosylate in tetrahydrofuran as solvent (Timko & Cram, 1974; Reinhoudt & Gray, 1975). Structure (173) is a representative example of a considerable number of other macrocyclic systems incorporating heterocyclic donor groups in their ring structure (Newkome, Sauer, Roper & Hager, 1977).

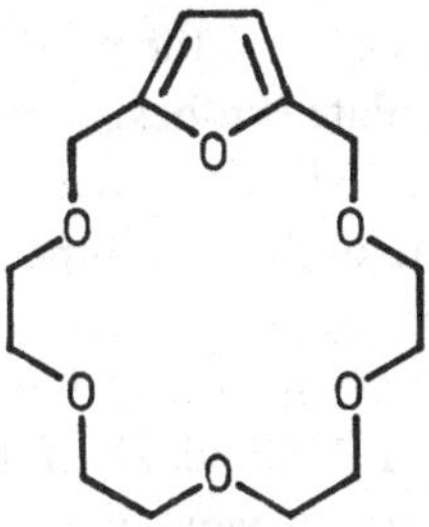

(173)

Effects of chain length on cyclization. The effect of chain length on cyclization of a series of acyclic precursors of benzo-crowns, of which (168) is the first member, has been studied (Mattice & Newkome, 1982). In accordance with kinetic data, calculations of the Monte Carlo type indicate that the probability of adoption of a conformation with a small 'head-to-tail' distance makes a significant contribution to the cyclization rate for the small to medium rings.

Template contributions. Alkali metal ions have been documented to play a template role in a number of crown syntheses. Thus, for example, the presence of K^+ has been shown to promote the formation of 18-crown-6 in syntheses such as [4.2] (Green, 1972); intermediates of type (174) are

(174)

probably involved. As in this example, the ideal template metal in such procedures appears, once again, to be one that best fits the cavity of the cyclic product. The small Li^+ has been used to aid formation of 12-crown-4 while larger Na^+ was used for 15-crown-5 (Cook *et al.*, 1974).

Kinetic aspects of the use of alkali metals as templates for the formation of other crowns have been studied in some depth (Mandolini & Masci, 1984). The results of such investigations parallel the previous observations – namely, that the catalytic efficiency of such ions in promoting cyclization shows a strong tendency to parallel their strength of binding with the crown products (this in turn often correlates with the fit of the metal ion for the macrocyclic cavity in the product).

In another category of template synthesis, ethylene oxide is cyclized in the presence of BF_3 and alkali, alkaline earth, or transition ions as their fluoroborate, fluorophosphate or fluoroantimonate salts – see [4.4]. This procedure yields cyclic tetramers, pentamers and/or hexamers with the particular product(s) obtained being quite dependent on the metal present. Thus $Ca(BF_4)_2$ gives a 50% yield of tetramer, $Cu(BF_4)_2$ and $Zn(BF_4)_2$ give the pentamer in 90% yield, whereas $Rb(BF_4)$ and $Cs(BF_4)$ yield exclusively hexamer (Dale & Daasvatn, 1976).

$$m \; CH_2\!\!-\!\!CH_2 \xrightarrow[\;M^{n+}\;]{\;BF_3\;} (CH_2CH_2O)_m \qquad [4.4]$$

$$m = 4,5,6$$

Further comments. The preceding discussion outlines typical syntheses for simple polyether crown rings. It needs to be noted that a considerable number of other types of crown derivatives, displaying a variety of molecular architectures, has also been synthesized. Many of these types parallel the 'structurally developed' macrocycles (which incorporate mainly donor atom types other than ether oxygen) discussed in Chapter

3. For example, the crown derivatives include ring systems containing appended side-chains incorporating additional binding sites, rigid polyether macrocycles whose cavities are preformed before complexation occurs, bi- and multi-nucleating systems capable of binding simultaneously to more than one metal ion, crowns incorporating attached colour and/or photoresponsive chromophores, as well as crowns immobilized on polymeric substrates. A further separate category involves the wide range of mixed-donor macrocycles which incorporate N, S, P or As donor atoms (as well as ether oxygens) in their rings. Aspects of the chemistry of selected examples from most of these ligand categories are discussed in the following sections.

Some crown-ether metal complexes

Metal-ion binding properties. Pedersen's paper marked the beginning of a new and major development in the study of the metal-ion chemistry of macrocyclic ligands. The crowns exhibit a number of unusual properties such as the ability to form complexes with alkali metal ions which, in certain cases, are stable in aqueous solution. Indeed, they also yield stable complexes of a range of non-transition metal ions but tend to bind less strongly to transition metal ions. Pedersen showed that rings incorporating between five and ten oxygen atoms tended to form the most stable complexes and complexes with some or all of the following metal ions were isolated: Li^+, Na^+, K^+, Rb^+, Cs^+, Ag^+, Au^+, Ca^{2+}, Sr^{2+}, Ba^{2+}, Cd^{2+}, Hg^+, Hg^{2+}, La^{3+}, Tl^+, Ce^{3+}, and Pb^{2+}. In such compounds the ether-oxygen interactions with the metal cation were considered to be essentially electrostatic.

Since this original work many new crowns have been synthesized and their complex formation has been very extensively studied. Complexes with many other ions (including weak complexes with several transition ions) have been characterized. Crystalline 1:1 (metal:ligand), 1:2, and 2:3 complexes as well as species of other stoichiometry have all been isolated.

X-ray structural studies. A considerable number of such complexes have been investigated by X-ray diffraction. Collectively, the structures are characterized by very wide diversity, and range from 'simple' coordination geometries through to quite complex ones (Truter, 1973; Hilgenfeld & Saenger, 1982).

One group of structures is characterized by the presence of a reasonable match of the metal ion for the radius of the cavity in the crown. An example of this type is given by the structure of the 1:1 complex between dibenzo-14-crown-4 and LiSCN (Shoham, Lipscomb & Olsher, 1983a). In this species, the Li^+ is five-coordinate, being bound to the four ether donors and to the nitrogen of the NCS⁻ ion in a nearly square pyramidal geometry. As expected from electrostatic considerations, the metal lies 0.79 Å out of the plane of the oxygens towards the axial thiocyanate ligand. A related structure occurs in the corresponding LiNCS complex of the slightly smaller ring, benzo-13-crown-4 (Shoham, Lipscomb & Olsher, 1983b). With the larger Rb^+ ion, a seven-coordinate complex is obtained when the crown is dibenzo-18-crown-6 (Bright & Truter, 1970). In this case, the large crown provides a basal planar array of six oxygen donors which lie approximately equidistant from the rubidium (Figure 4.1). Axial coordination of a thiocyanate ion again occurs to complete the coordination geometry.

A second group of structures results when the cavity of the crown, in a non-folded configuration, is too large relative to the radius of the metal ion. In such a case, the crown will normally wrap around the cation (unless some of the oxygens do not coordinate). Alternatively, in special cases, the crown may coordinate two metal ions in its cavity (see later). An example of a 'wraparound' arrangement is given by the 1:1 complex between K^+ and dibenzo-30-crown-10 (172). On complexation, the ligand is not flat but folds around the K^+ ion such that all ten oxygens coordinate (Figure 4.2) – the iodide ion remains uncoordinated (Bush & Truter, 1972). By folding, the macrocycle produces a smaller (three-dimensional) cavity which is more suited to the steric and electronic requirements of the K^+ ion.

The final group of complexes arises when the metal ion is too large to fit inside the available crown cavity. There is a tendency in this situation for

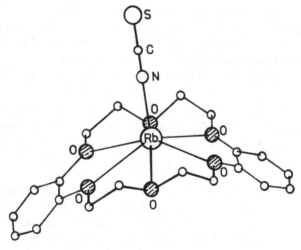

Figure 4.1. X-ray structure of the RbNCS complex of dibenzo-18-crown-6 (Bright & Truter, 1970).

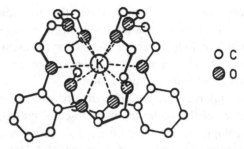

Figure 4.2. X-ray structure of the cationic complex of potassium with dibenzo-30-crown-10 (Busch & Truter, 1972).

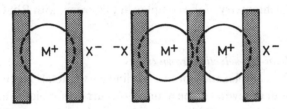

Figure 4.3. Typical arrangements for crown 'sandwich' complexes.

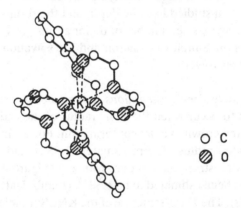

Figure 4.4. Structure of the bis-ligand potassium complex of benzo-15-crown-5. The complex shown is cationic (Mallinson & Truter, 1972).

complexes to form which exhibit ligand-to-metal ratios of greater than 1:1. In a number of cases, compounds showing stoichiometries of 1:2 and 2:3 (M:L) occur. Typically, sandwich structures of the types illustrated in Figure 4.3 occur. An example of the former type is given by the 1:2 complex of potassium iodide with benzo-15-crown-5 (Mallinson & Truter, 1972). The K^+ ion lies at the centre of symmetry of the sandwich (Figure 4.4) such that the ten oxygen donors from the two ligand molecules define a pentagonal anti-prismatic coordination geometry.

Further considerations. It needs to be noted that the stoichiometry of a given crown-metal complex is not only influenced by ring size; a range of other factors which include the charge density on the metal, the nature of the anion, and the relative strain energies of the crown in different conformations may all make a contribution.

Similarly, the apparent stoichiometry of a crown complex in the solid is not necessarily a reliable measure of the actual metal-ligand binding situation. The complex of rubidium thiocyanate with dibenzo-18-crown-6 has an apparent metal:ligand ratio of 2:3 in the solid. However, the X-ray structure reveals that, in fact, the complex crystallizes such that one in three of the crown molecules in the lattice do not contain a Rb^+ ion. Thus the true stoichiometry of the complex of this crown with RbSCN is 1:1.

Metal-ion-selective crown ethers

The complexation of the alkaline earth metals is reminiscent of the behaviour of several of the naturally occurring antibiotics and, like the latter, the crown often exhibits remarkable selectivity for particular ions. The thermodynamic factors underlying the selectivity of many of the crowns have been studied in some depth and the results related to such parameters as cavity size, number of donor atoms present, possible ring conformations on complex formation and the solvation energies of the various species involved.

Macrocyclic cavity size considerations. For small ring crowns, enhanced binding tends to occur when the ionic radius of the cation matches the cavity size of the crown in a flat conformation. For example, 18-crown-6 has an estimated radius of approximately 1.38 Å and, with the alkali metals, forms its strongest complex with K^+ (Figure 4.5) whose ionic radius has also been estimated to be 1.38 Å (Lamb, Izatt, Christensen & Eatough, 1979). The X-ray structure of the KNCS complex of 18-crown-6 confirms that the K^+ occupies the macrocyclic cavity in an unstrained manner with the ether oxygens arranged regularly around this cation in a near planar fashion (Seiler, Dobler & Dunitz, 1974). With Na^+, the X-ray structure of the NaNCS complex of 18-crown-6 appears to reflect the reason for the drop in stability relative to the K^+ species. In this complex, the coordinated 18-crown-6 is bent such that one donor oxygen lies well out of the approximate plane formed by the other five (Dobler, Dunitz & Seiler, 1974). This non-planarity of the donor set appears to be a consequence of the less than ideal fit of the metal for the macrocyclic cavity of 18-crown-6 in a planar conformation. Nevertheless, the origins

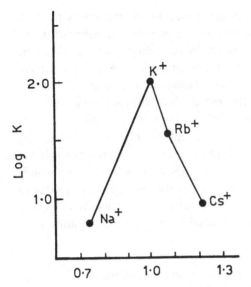

Figure 4.5. The selectivity of 18-crown-6 for potassium ion (in water at 25 °C).

of the stability pattern illustrated in Figure 4.5 may not be quite as straightforward as suggested by simple hole-size considerations since there is evidence that solvation effects also markedly influence the relative stabilities of the crown complexes of these ions (Izatt *et al.*, 1985). Indeed, it has been demonstrated that the observed stability constant sequence shown in Figure 4.5 for 18-crown-6 (or the corresponding sequences for its benzo or dibenzo derivatives) may vary in other solvents (Wong, Konizer & Smid, 1970).

Cavity size effects of the type discussed above are no longer of major importance when considering the complexing ability of the larger crowns. For these, the requirement for full participation of all the available oxygen donors on complexation may be of greater significance. Full participation will result in the occurrence of maximum electrostatic interaction between the crown and the metal cation and hence will be expected to result in a more favourable enthalpy term for complex formation. It is only for the smaller ring crowns that the most favourable enthalpy of complexation will normally correspond to a close match of the radius of the metal ion for the hole size of the macrocyclic cavity. Nevertheless, it is sometimes found that individual large-ring crowns tend to show selectivity for larger ions (such as caesium) relative to other, smaller ions. This property may largely reflect the greater ease with which the large ion is able to accommodate the increased number of donor oxygens present in such large rings. In one sense, such rings can be considered to provide a 'synthetic' replacement for the 'natural' solvent shell surrounding the free metal ion. However, it is stressed that it is difficult to generalize about the metal-ion affinities shown by many crown ligands – clearly, as is evident from the previous discussion, the 'match of the metal to the hole' concept is only of limited utility. Nevertheless, the selectivity that is observed for many crowns has led to many applications in analytical chemistry (Koltoff, 1979). These include the use of crowns as reagents in solvent extraction, the development of new ion-selective electrodes, and the attachment of crowns to polymeric supports to yield chromatographic materials for metal-ion separation.

Mixed-donor crown ethers

Some representative examples. Many crown macrocycles incorporating other heteroatom types besides ether oxygens have been synthesized. In an early preparation of this type, Lehn *et al.* reported a synthesis for the O_4N_2-system (175, diaza-18-crown-6) using the procedure outlined by [4.5] (Dietrich, Lehn & Sauvage, 1969).

Aromatic azacrowns have also been synthesized. For example, condensation of *o*-hydroxyaniline with the appropriate dichloro polyether yields benzo-12-azacrown-4 or benzo-15-azacrown-5 (176) (Lockhart *et al.*, 1973). Starting from 2,6-dibromopyridine, a number of pyridine-containing crowns such as (177) have also been obtained (Timko *et al.*, 1974; Vogtle & Weber, 1974).

(175)

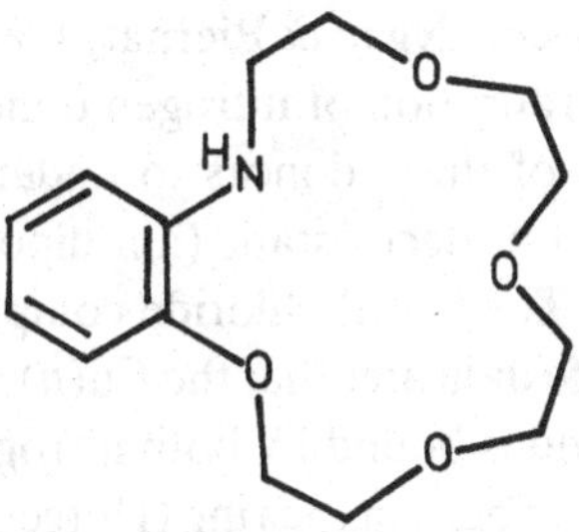

(176)

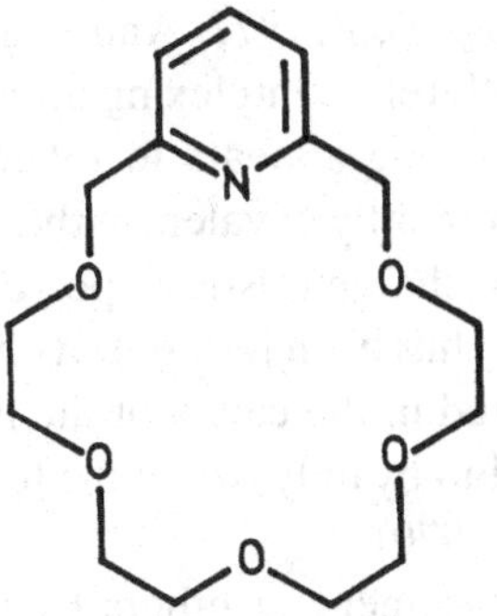

(177)

A variety of thia-derivatives in which thioether groups replace ether groups have also been synthesized. For instance, using a high-dilution procedure, (178) and a number of its analogues containing other ring sizes have been prepared by cyclization of thioglycol or the appropriate polyether dithiol with a dichloropolyether in ethanol in the presence of sodium hydroxide (Bradshaw *et al.*, 1976).

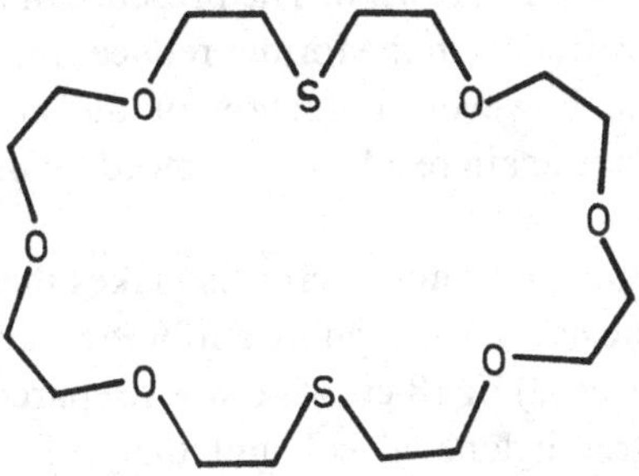

(178)

Effect on complex stability. Structurally, mixed donor crowns may be considered to lie between the two (extreme) major macrocyclic categories discussed so far and, in general, their properties also tend to span those of each category. Thus, there is more tendency for such rings to form stable complexes with transition and other heavy metal ions while their complexes with the alkali metals become less stable (relative to the corresponding rings containing only ether donors). For example, the log of the binding constant for the diaza derivative (175) with K^+ in methanol is 2.0 whereas the analogous value for 18-crown-6 is 6.1. The corresponding log values for complexation of Ag(I) in water are 7.8 and 1.6, respectively (Frensdorff, 1971).

Solution measurements have confirmed that (175) forms moderately stable complexes with a number of transition ions (Arnaud-Neu, Spiess,

Schwing-Weill, 1977; Anderegg, 1981; Luboch, Cygan & Biernat, 1983). The different complexing affinities on incorporation of nitrogen donors into the ring appears to reflect the ability of these donors to undergo predominantly covalent rather than the mainly electrostatic (ion-dipole) binding characteristic of polyether groups. The Cu(II) chloride complex of (175) has been isolated. Its X-ray structure indicates that the Cu(II) ion is located in the cavity of the macrocycle and is bound by both nitrogen atoms but by only two of the four available oxygens in the ring (Herceg & Weiss, 1970).

Replacement of ethers by thioethers in crown compounds [see, for example (178)] also reduces their affinity for the alkali metals and again leads to a tendency to complex heavy metals such as Ag(I) more strongly (Pedersen, 1971; Frensdorff, 1971).

Substituted crown polyethers

The effect of ring substituents. The cation selectivity of a given crown may be altered by variation of the ring substituents. For example, the dibenzo-18-crown-6 molecule is a generally poorer ligand than its unsubstituted derivative, 18-crown-6. The presence of the electron-withdrawing benzene substituents in the former reduces the donor capacity of nearby ether groups; however, reduction of the aromatic groups to cyclohexanyl rings once again results in enhanced binding power for this latter product.

Attachment of carbonyl groups to crowns makes these products more akin structurally to the natural ionophore antibiotics such as valinomycin. The dioxo-derivative (179) of 18-crown-6 was prepared in 35% yield by condensation of tetraethylene glycol and diglycolic acid chloride in benzene at 50 °C for 48 hours (Izatt *et al.*, 1977a and 1977b). This product gives binding constants for Na^+, K^+ and Ba^{2+} in methanol which are 10^2–10^4 times less stable than for the parent crown – the lower constants are a reflection of less favourable ΔH values for complexation in these

(179)

cases. In this study it was found that the stability constants for both (179) and 18-crown-6 fall in the order $Ba^{2+} > K^+$ whereas the reverse order occurs for valinomycin.

Ring substituents incorporating a donor atom. In a study of the effects of ligand substituents containing a donor atom on complex formation, 1,3-xylyl derivatives of type (180) were prepared. The functional groups (–R) are 'inwards-orientated' such that, when suitable donors are present, they are able to act as further coordination sites (Newcomb & Cram, 1975; Koenig, Helgeson & Cram, 1976). The anionic forms of the carboxylic acid derivatives of the 15-, 18-, 21- and 30-membered rings were tested for their ability to transport Li^+, Na^+, K^+ and Ca^{2+} from water into dichloromethane. The distribution of a cation between these two solvents was found to be ring-size dependent: the 18-membered species was best for Li^+, the 21-membered for Na^+, the 30-membered for K^+ while, for Ca^{2+}, the 18-membered ring was again the most efficient carrier.

R = H, Cl, Br, CN
COOH
CH$_2$OH
COOCH$_3$
CH$_2$OCH$_3$
OCH$_3$

(180)

Species such as (180) form a category of crowns incorporating an additional donor function which is not directly part of the macrocyclic ring. Many other derivatives have been synthesized in which additional donor functions are appended to the crown ring by means of 'arms'. A selection of such pendant arm crowns is discussed in the following section.

Crown polyethers containing pendant donor groups
The lariat crown ethers. A number of 'lariat' crown ethers which possess additional binding sites on flexible chains attached to the crown ring have been synthesized (Schultz, Dishong & Gokel, 1982). Structurally, these species are related to both the cyclic 'coronand' and open-chain ligand categories. Their complexes usually show additional stability relative to those of the simple crowns.

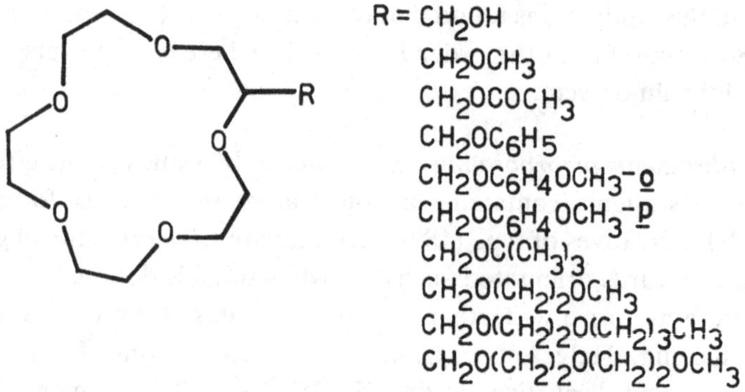

$$R = CH_2OH$$
$$CH_2OCH_3$$
$$CH_2OCOCH_3$$
$$CH_2OC_6H_5$$
$$CH_2OC_6H_4OCH_3\text{-}\underline{o}$$
$$CH_2OC_6H_4OCH_3\text{-}\underline{p}$$
$$CH_2OC(CH_3)_3$$
$$CH_2O(CH_2)_2OCH_3$$
$$CH_2O(CH_2)_2O(CH_2)_3CH_3$$
$$CH_2O(CH_2)_2O(CH_2)_2OCH_3$$

(181)

Typical lariat ethers are illustrated by the general structure (181).

The extraction ability (and selectivity) for Na^+/K^+ (in CH_2Cl_2/water) of this ligand series has been investigated in the presence of picrate anions. Enhanced extraction was found to occur when the pendant group has donor atoms which are correctly orientated for axial interaction with a metal bound in the macrocyclic portion of the molecule. For instance, the extraction of Na^+ is enhanced by a factor of two when R in (181) is the *ortho* derivative of $-CH_2OC_6H_4OCH_3$ rather than the *para* one. The former is able to align its methoxy group for axial interaction with the Na^+ whereas the latter cannot. The enhanced extractability in the former case is undoubtedly a consequence of both increased neutralization of charge on the alkali ion and the additional lipophilicity induced by the shielding of the ion by coordination of the pendant group.

The series of derivatives of diaza-18-crown-6 illustrated by (182) have also been used for extraction studies involving Li^+, Na^+ and K^+ picrates in a CH_2Cl_2/water system (Cho & Chang, 1980). Relative to the parent macrocycle, all the N,N'-derivatives led to enhanced extraction of these ions although the selectivities observed for the parent system were moderated somewhat.

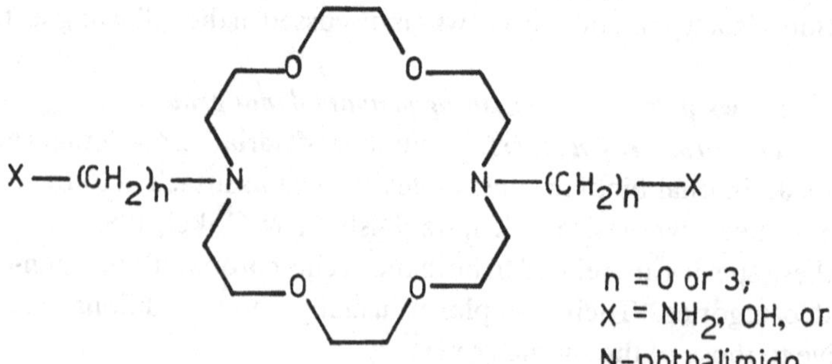

$$X-(CH_2)_n-N \qquad N-(CH_2)_n-X$$

n = 0 or 3;
X = NH_2, OH, or
N-phthalimido

(182)

(183)

Other pendant arm systems. Many other pendant arm crowns have been synthesized – one system studied in some detail is given by (183) where R is a range of aliphatic derivatives incorporating further oxygen donors (Masuyama, Nakatsuji, Ikeda & Okahara, 1981; Kaifer *et al.*, 1982; Davidson *et al.*, 1984).

Once again, such species yield Na^+ and K^+ complexes which are more stable than complexes of their unsubstituted analogues. This is illustrated in Table 4.1 for the case where $n = 2$: the Table summarizes $\log K$ values

Table 4.1. Log K values for Na^+ and K^+ complexes of ligands of type (183; $n = 2$) in methanol[a]

Complexes with R = $-(CH_2CH_2O)_yH$

Metal	$y = 0$	$y = 1$	$y = 2$	$y = 3$
Na^+	2.8	4.8	4.3	4.3
K^+	4.2	5.5	5.9	5.7

Complexes with R = $-(CH_2CH_2O)_yCH_3$

Na^+		5.5	5.7	4.3
K^+		5.4	–	6.0

[a]Masuyama, Nakatsuji, Ikeda & Okahara, 1981; Davidson *et al.*, 1984.

for the Na^+ and K^+ complexes of two ligand series based on the aza-18-crown-6 parent.

Solubility and crown complexation

As is implied in the previous discussions, marked changes in solubility usually accompany complexation involving the crowns. Thus these ligands in the presence of salts may result in a mutual increase in solubility of both the ligand and the salt owing to complex formation. Sometimes such effects are very considerable, for instance, the solubility of dibenzo-18-crown-6 in methanol is 1×10^{-3} mol dm^{-3} whereas the solubility of the potassium thiocyanate complex is a hundredfold greater at 1.07×10^{-1} mol dm^{-3} (Pedersen, 1967). Even though many simple crown ethers are moderately soluble in water, when complexed the hydrophilic interior of such molecules is masked. Thus, surrounding a Lewis acid such as Na^+ or K^+ with a crown confers an increase in the lipophilicity of the system. The concomitant greater solubility of the complexed cation (as its ion pair with the counter ion) in non-polar solvents has led to numerous applications in both analytical and synthetic organic chemistry.

Crowns as extraction reagents

As discussed already, crowns may be involved in solvent extraction processes in which an inorganic reagent is transferred (sometimes selectively) from one phase (often water) into an immiscible organic phase; the extraction involves ion-pair formation between the cationic crown complex and the counter ion (Blasius & Janzen, 1981).

The ease of extraction may be represented by the extraction coefficient (K_{ex}) for the system:

$$M^{m+}_{(aq)} + C_{(org)} + mA^{-}_{(aq)} \overset{K_{ex}}{\rightleftharpoons} (MCA_m)_{(org)}$$

$$\text{where} \quad K_{ex} = \frac{[MCA_m]_{(org)}}{[M^{m+}]_{(aq)}[C]_{(org)}[A^-]^m_{(aq)}}$$

MCA_m is the ion-pair formed between the crown (C) complex containing the metal ion (M^{m+}) and A^- is the counter ion. It needs to be noted that the degree of extraction is anion-dependent. For example, the extraction of an alkali metal into an organic phase is enhanced when the counter ion is a large anion such as picrate. Alkali metal picrates undergo extraction into benzene in the presence of 18-crown-6 in the order $K^+ > Rb^+ > Cs^+ > Na^+$ (Iwachido, Sadakane & Toei, 1978). Divalent ions may also be extracted. For 15-crown-5 in benzene, the picrate extraction coefficients (from water) fall in the order $Pb^{2+} > Sr^{2+} > Ba^{2+} > Ca^{2+}$

(Takeda & Kato, 1979). An interesting analytical application involving dicyclohexanyl-18-crown-6 in chloroform is the separation of Sr^{2+} from Ca^{2+} in milk; the procedure may be applied to the estimation of $^{89,90}Sr^{2+}$ in milk (Kimura, Iwashima & Hamada, 1979).

Apart from analytical applications, reports of the use of crowns in synthetic organic chemistry have been quite common. Typically, the solubilization of an inorganic reagent (such as potassium permanganate) or the production of a 'free' counter ion (such as the fluoride ion) in an organic solvent such as benzene has formed the basis for many of these reports.

Crown ethers in synthetic organic chemistry

An area in which crowns (and their three-dimensional analogues – the cryptands) find wide application is that of synthetic organic chemistry. Reactions involving the use of polyether macrocycles include the following: saponification, esterification, redox reactions, nucleophilic substitution (fluorination, alkoxylation, cyanation, nitration, etc.), elimination (carbene and nitrene formation, decarboxylation, etc.), condensation (alkylation, arylation, etc.), rearrangements (Cope, Wagner–Meerwein, Smiles, Claisen, etc.), Wittig reactions, Cannizzaro reactions, Michael additions as well as a number of other synthetic procedures (Knipe, 1976; Weber & Gokel, 1977; Hiraoka, 1982). Because of the diversity of such reaction procedures, no attempt will be made here to cover this large field of activity. Rather, a few illustrative examples are described.

Solubilization and generation of naked anions. The use of crowns in synthetic organic chemistry relies on the inherent chemical stability of the rings themselves towards substitution, elimination and cleavage reactions plus, as mentioned already, their capacity to enhance the solubilities of a range of metal salts in organic solvents. Their success as useful reagents reflects the ability of the crowns to complex the cation of an alkali metal salt in an aprotic solvent and hence moderate, to a greater or lesser degree, the formation of a tight ion pair (between the metal ion and the counter ion present). The consequent reduction in polarity of the system is, in part, the origin of the solubilization of such reagents in organic media. For example, when the counter ion is the fluoride ion, this ion's effectiveness as a nucleophile (and as a base) may be increased. Under these conditions, the fluoride has been termed a 'naked' anion and may be considered to be, at best, only weakly solvated. As a consequence, both the higher charge density (with resultant increased base strength) and the smaller effective size (with corresponding reduction in steric hindrance effects) will enhance its capacity to act as a nucleophile.

It should be noted that the three-dimensional polyether cages (the cryptands) are usually most effective at producing 'naked anions'. With these, the metal ion is completely encapsulated by the polyether network and thus better charge separation is achieved. In the case of the crowns, such complete encapsulation does not normally occur and hence the counter anion is more readily able to associate directly with the complexed metal cation. In such cases, the use of the term 'naked' is somewhat of a misnomer.

Representative organic syntheses. A solution of the 'naked' fluoride ion may be generated by dissociation of KF in an acetonitrile or benzene solution containing 18-crown-6 (Liotta & Harris, 1974). The considerable nucleophilicity of this anion under these conditions is demonstrated by the fact that it is capable of displacing leaving groups from both sp^2 and sp^3 hybridized carbons in a number of structural environments.

A range of other active anions may be generated using a similar procedure – for example, KBr, KI, $KOCH_3$, KCN and $KOOCCH_3$ have all been used for this purpose. Thus, the complexed form of $KOCH_3$ reacts with *o*- and *m*-dichlorobenzene at 90 °C to yield the corresponding chloroanisole. The general reaction provides a rare example of nucleophilic aromatic substitution (by the methoxide ion) of an unactivated aromatic halide (Sam & Simmons, 1974).

Using dicyclohexyl-18-crown-6 it is possible to dissolve potassium hydroxide in benzene at a concentration which exceeds 0.15 mol dm^{-3} (Pedersen, 1967). The free OH$^-$ has been shown to be an excellent reagent for ester hydrolysis under such conditions. The related solubilization of potassium permanganate in benzene, to yield 'purple benzene', enables oxidations to be performed in this solvent (Hiraoka, 1982). Thus, it is possible to oxidize a range of alkenes, alcohols, aldehydes, and alkylbenzenes under mild conditions using this solubilized reagent. For example, 'purple benzene' will oxidize many alkenes or alcohols virtually instantaneously at room temperature to yield the corresponding carboxylic acids in near-quantitative yields (Sam & Simmons, 1972).

Use of dibenzo-18-crown-6 together with sodium borohydride also enhances the reducing power of this reagent when reacting with a number of ketones (to yield the corresponding alcohols) (Matsuda & Koida, 1973). Once again, the effect presumably reflects the generation of a naked BH_4^- ion in these cases.

Phase transfer catalysis. As well as their use in homogeneous reactions of the type just described, polyethers (crowns and cryptands) may be used to catalyse reactions between reagents contained in two different phases (either liquid/liquid or solid/liquid). For these, the polyether is present in only 'catalytic' amounts and the process is termed 'phase transfer catalysis'. The efficiency of such a process depends upon a number of factors. Two important ones are: the stability constant of the polyether complex being transported and the lipophilicity of the polyether catalyst used.

It needs to be noted that phase transfer catalysis has implications for energy conservation; for example, reactions which normally require heat may proceed at room temperature in the presence of naked anions.

A typical phase transfer catalytic reaction of the liquid/liquid type is the cyanation of an alkyl halide in an organic phase using sodium or potassium cyanide in an aqueous phase. When these phases are stirred and heated together very little reaction occurs. However, addition of a small amount of crown ether (or cryptand) results in the reaction occurring to yield the required nitrile. The crown serves to transport the cyanide ion, as its ion pair with the complexed potassium cation, into the organic phase allowing the reaction to proceed.

Other liquid-ligand two-phase reactions mediated by polyethers include anion promoted C-alkylations, oxidations, and (borohydride) reductions. In such cases, the organic substrate and a catalytic amount of polyether in an organic phase are shaken with a saturated aqueous solution of the required anionic reagent.

In phase transfer catalysis of the solid/liquid type, the organic phase (containing dissolved organic reactant and a small amount of the crown) is mixed directly with the solid inorganic salt. Such a procedure enables the reaction to proceed under anhydrous conditions; this is a distinct advantage, for example, when hydrolysis is a possible competing reaction. Because of their 'open' structure, crown ethers are readily able to abstract cations from a crystalline solid and are often the catalysts of choice for many solid/liquid phase transfer reactions.

Crown ethers attached to insoluble polymeric substrates (see the following discussion for examples) have been used as phase transfer catalysts for liquid/liquid systems. In using such systems, the catalyst forms a third insoluble phase; the procedure being referred to as 'triphase catalysis' (Regen, 1979). This arrangement has the advantage that, on completion of the reaction, the catalyst may be readily separated from the reaction solution and recycled (Montanari, Landini & Rolla, 1982). As such, the facile separation may considerably aid the purification of the required organic product.

Immobilized crown ethers

Typical systems. A considerable number of immobilized polyether systems have been synthesized both for phase transfer catalysis as just discussed and for use in a number of analytical applications. Such immobilized systems are generally synthesized by either copolymeriz-ation of suitably functionalized macrocycles in the presence of cross-link-ing agents or by appending functionalized macrocycles to existing polymeric substrates. Structures (184)–(186) give examples of different

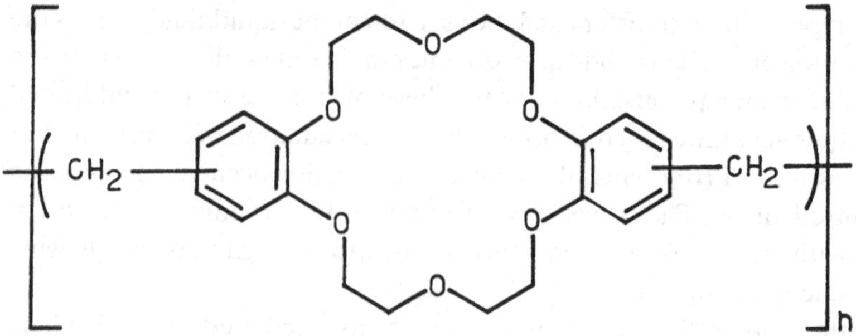

(184)

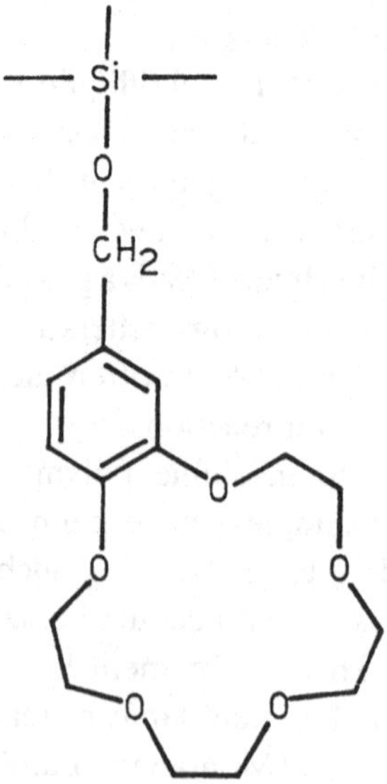

(185)

(186)

polymer types prepared by such means (Blasius & Janzen, 1982). Attention has been directed to the use of polymeric species of this type as ion-selective membranes, as chromatographic stationary phases (including their use in high performance liquid chromatography, HPLC) and even for trace enrichment of radionuclides.

For (184)–(186), since these are neutral ligands the uptake of a metal cation is also accompanied by anion binding such that electro-neutrality is maintained. Thus resins of this type may sometimes be useful for separation of either cations or anions.

Polymer (184) has a network structure and was obtained by reaction of dibenzo-18-crown-6 with formaldehyde in formic acid. Amongst the alkali metal ions, it selectively captures K^+ and Cs^+ from methanol or methanol/water. A related polymeric product has been reported (as a gel) from the reaction of this crown with formaldehyde in chloroform using sulfuric acid as catalyst (Davydova, Baravanov, Apymova & Prata, 1975).

The immobilized crown system (185) uses a silica gel substrate: attachment to the silica is possible by using the reactivity of the silanol groups on its surface. This system has been used to separate F^-, Cl^-, Br^- and I^-, from each other as well as SO_4^{2-} from Cl^-. These separations were achieved in aqueous solution using Na^+ as the counter ion. Effective separation of SO_4^{2-} and Cl^- anions was achieved for solutions in which the molar ratio $SO_4^{2-}:Cl^-$ was up to 1:10 000 (Blasius & Janzen, 1982).

Starting from the corresponding hydroxymethyl-benzocrown, it has been possible to generate the immobilized system (186) by reacting the above precursor with chloromethylated polystyrene (which is available commercially as Merrifield's resin). Typically, systems of this type contain a polystyrene matrix which has been cross-linked with approximately 1–4% p-divinylbenzene. In one study involving (186), a clean resolution of the alkali metal halides was achieved by HPLC using (186) as the solid phase and methanol as eluent (Blasius *et al.*, 1980). In other studies, the divalent alkaline earths were also separated.

Aza-crowns have also been appended to polystyrene – in some cases involving a direct reaction between the chloromethylene groups of the starting polymer and a secondary amine function on the macrocycle. In other cases, attachment occurred, via a spacer group, to another part of the molecule. Once again, such species have been used to separate alkali and alkaline earth metals (and also transition metals in a few studies) using column chromatography. For many of the crown and aza-crown (cross-linked) polystyrene systems, binding of metal ions seems to be somewhat enhanced relative to the free crowns. However, with these systems, the separations tend to be limited by the low capacity of the resins and also, in some instances, by the tendency for slow diffusion of the metal cations within the resin matrix.

Some further comments. A large number of other immobilized polyether systems of the above general type have now been synthesized containing a variety of crowns and attachment linkages. For example, 'spacer' groups of different lengths have been used to vary the distance between the crown and polymeric backbone. A range of other properties may also be varied. These include: the swelling ability of the polymer, its degree of functionalization and of cross-linking, its hydrophobic or hydrophilic nature, and the ease of accessibility of metal ions for the binding sites. By varying the above parameters, it is in theory possible to design 'tailor made' derivatives for particular applications. Success in this area has already been achieved in a few instances; however, the development of new polymeric systems having predetermined properties will un-doubtedly continue to provide a considerable challenge in the future.

Chromogenic crown ethers

A range of crown derivatives have been developed in which colour-inducing functional groups are incorporated in the overall structure of the molecule. These reagents are designed to give rise to specific colour changes on complexation of normally 'colourless' metal ions such as the alkali and alkaline earth cations. A range of derivatives of this type have been developed for use as spectrophotometric analytical reagents for particular cations – some specificity may be built into each system by choosing the appropriate cavity size in the crown ether portion of the molecule.

The colours of these compounds are associated with charge-transfer transitions involving the dye moiety. Thus, when the steric and/or electronic effects of coordination cause variation in the energy of these transitions, then the induced spectral changes may be used to monitor complex formation.

Non-ionizable reagents. One group of chromophoric reagents is characterized by not containing ionizable protons in their structure. A very large number of interesting molecules of this type have been reported (Vogtle, 1980; Dix & Vogtle, 1981). Structures (187)–(189) give representative examples of such reagents. For many of these systems, it has been demonstrated that the modified rings retain the metal-ion discriminating ability which is characteristic of the corresponding simple crowns.

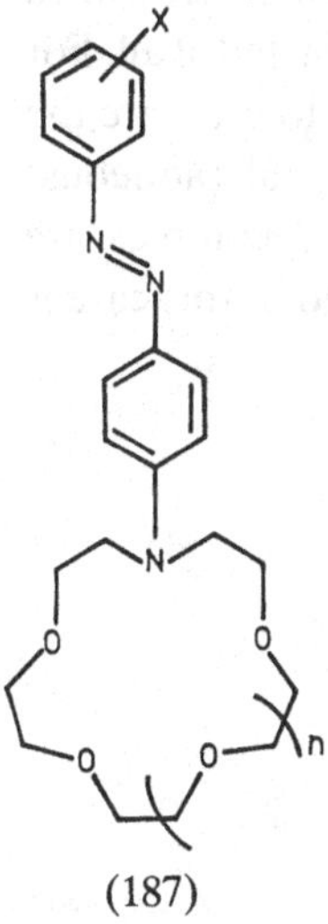

(187)

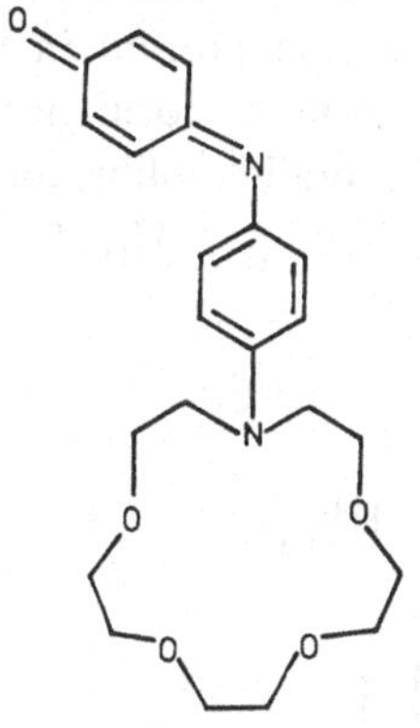

(188)

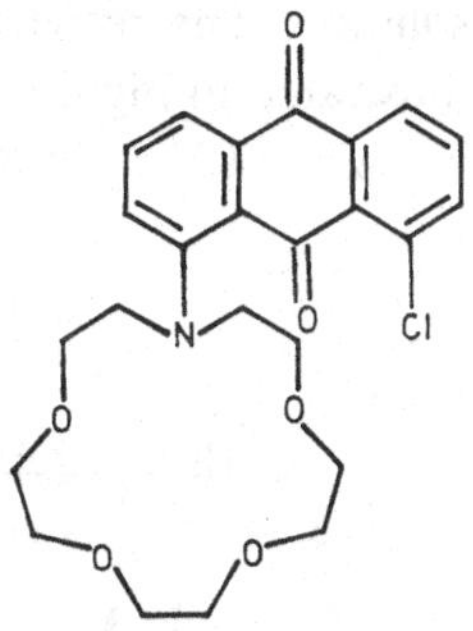

(189)

Ionizable reagents. The second category of chromogenic reagents are those incorporating a chromophoric group bearing an ionizable proton (or protons) (Takagi & Ueno, 1984); in such systems, the colour change is associated with ionization of the proton. The protonated dye group is incorporated in the structure near the crown ether skeleton such that

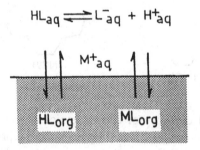

| X | H | NO₂ | Br | H | NO₂ |

(190)

deprotonation is assisted by complexation of the metal cation at the crown centre. In some cases the deprotonated dye moiety also contributes to the overall stability of the product by coordinating directly to the metal.

The first series of chromogenic crowns to be studied in depth had the general structure (190). The parent species (190; $X = H$, $n = 1$) is virtually insoluble in water at neutral pH but is soluble in aqueous alkali and in common organic solvents. In this system, it is the amino-proton which is acidic and its loss results in a colour change from orange to bright red. The alkali ions K^+ and Rb^+ are strongly extracted from basic aqueous solution into chloroform containing this crown derivative; the development of a bright red coloration in the chloroform phase confirms the presence of the alkali metal complex of the deprotonated crown. In contrast, Na^+ is only weakly extracted while Li^+ is not extracted at all. For the former ions, spectrophotometric means may be used to estimate the concentration of the particular ion present in the original (aqueous) solution. The respective equilibria for formation of a 1:1 complex are illustrated in Figure 4.6. The overall equilibrium involved in this case is given by: $LH_{(org)} + M_{(aq)}{}^+ \rightleftharpoons ML_{(org)} + H_{(aq)}{}^+$.

$$HL_{aq} \rightleftharpoons L^-_{aq} + H^+_{aq}$$

$$M^+_{aq}$$

$$HL_{org} \qquad ML_{org}$$

Figure 4.6. Equilibria underlying the extraction of the alkali metal ions by ligands of type (190).

This equilibrium may be used to define an extraction constant (K_{ex}) in the usual way. It should be noted that under some conditions, especially when larger ions are involved, a second equilibrium involving complexation of a further molecule of (non-deprotonated) ligand may also be present.

Variation of the X-substituent in (190) may be used to tune the pH at which efficient extraction occurs. For example, for the case where X = NO_2, the electron-withdrawing ability of this group results in more facile deprotonation of the amino proton. Thus the extraction may be performed at a lower pH than for the system with X = H. Similarly, variation of *n* in (190) may be used to tune the system for metal ions of different radii.

As an illustration of the application of reagents of the above type, (190; X = H, *n* = 1) has been used to estimate the concentration of K^+ in sea water in the presence of an approximate two-hundredfold molar excess of Na^+ ion.

A second group of reagents has the ionizable chromophoric group incorporated in a side arm attached to the crown ring such that, on complex formation, the deprotonated site is able to interact directly with the metal ion contained in the cavity of the crown. Such molecules are thus formally analogous to the 'pendant functional group' macrocycles discussed previously. Typical of this group is the aza-18-crown-6 derivative (191) incorporating a *p*-nitrophenol side arm (Nakamura, Sakka, Takagi & Ueno, 1981) and the related, all-oxygen donor species (192), derived from 15-crown-5 (Nakamura, Nishida, Takagi & Ueno, 1982). These systems selectively extract particular alkali ions from water into organic solvents (such as 1,2-dichloroethane) via species such as (193).

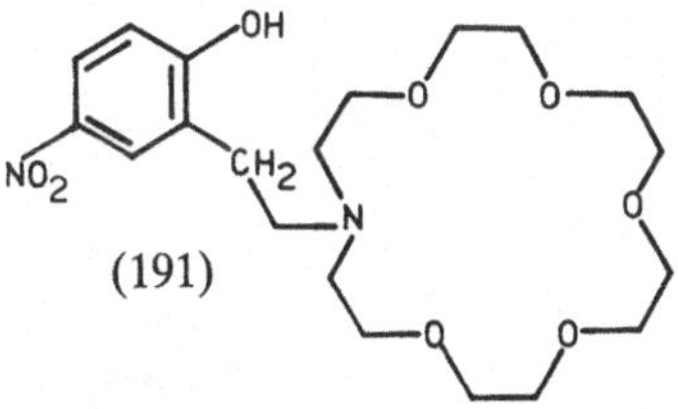

(191)

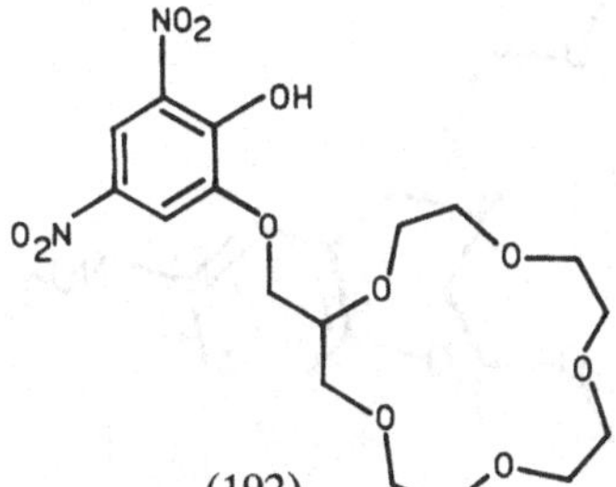

(192)

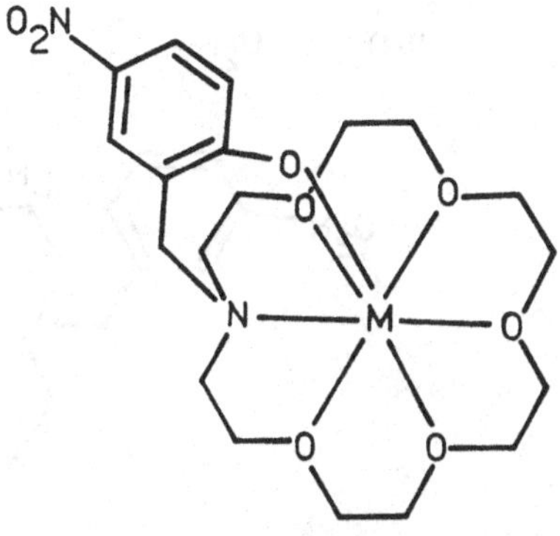

(193)

Unlike the previously mentioned systems, compounds such as (191) and (192) show no tendency to yield 2:1 (L:M) complexes. These compounds show much promise as analytical reagents for the alkali metals and, for instance, have been demonstrated to be suitable for the extraction and spectrophotometric determination of Na^+ in human blood serum.

A range of other chromogenic crowns have been studied including derivatives such as (194) incorporating a substituted azobenzene moiety and a phenol-containing crown ring (Kaneda, Sugihara, Kamiya & Misumi, 1981). This species has been used as a sensitive reagent for the spectrophotometric determination of Li^+. Diprotic systems have also been developed: these were designed to be selective for divalent ions and especially for the alkaline earths. Structure (195) (Nishida, Tazaki, Takagi & Ueno, 1981) represents one of a number of such reagents which, in general, show the expected favourable complexing ability towards many divalent metal ions. For example, with (195), the deprotonated phenol groups are sterically able to occupy axial coordination positions above and below a metal ion held in the cavity of the crown.

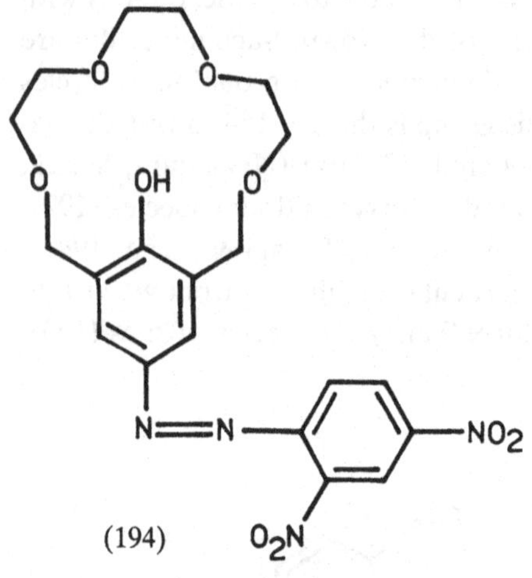

(194)

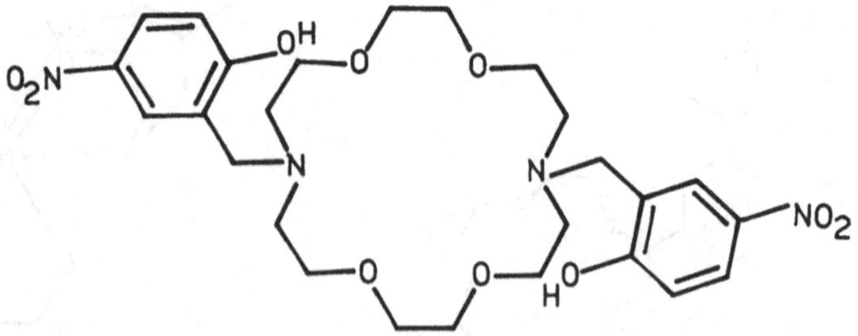

(195)

Photoresponsive crown ethers

Photoresponsive biochemical processes are found throughout nature and include photosynthesis, vision phototropism, and phototaxis. In an attempt to incorporate related photoresponsiveness in macrocyclic reagents, crown derivatives incorporating a photoresponsive functional group which is able to induce a structural change in the metal-binding site on irradiation have been synthesized. A prerequisite for a satisfactory photoresponsive system is that it exhibits a high quantum yield, has high reversibility, and involves a significant structural change. Processes such as the photo-induced dimerization of anthracene and the (E)–(Z) isomerism of azobenzene (Figure 4.7) both fulfill these criteria. This latter chromophore has formed the basis for a number of photoresponsive crown derivatives (Shinkai & Manabe, 1984) in which the crown portion of the molecule is covalently bound to the azobenzene chromophore. On reaction with a photon, the induced structural charge is transmitted to the crown polyether portion of the molecule such that the coordination properties of the latter are modified. The system thus contains an 'optical switch' which may serve to control the complexation/decomplexation cycle.

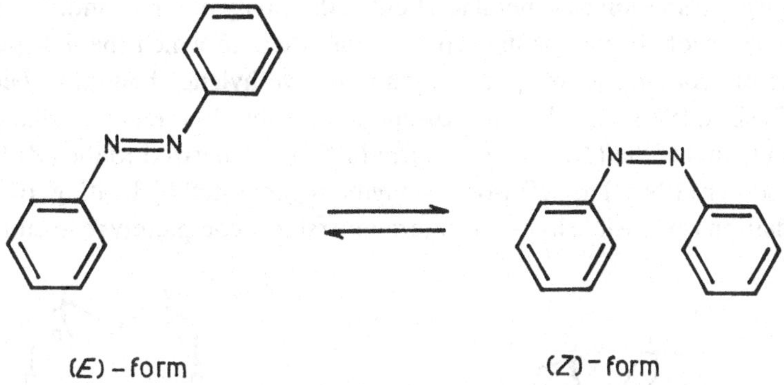

Figure 4.7. The (E)–(Z) isomerism of azobenzene.

Photoresponsive systems incorporating an azobenzene moiety. The capped crown ether (196), shown as the (E) isomer, was synthesized initially by a high-dilution condensation between diaza-18-crown-6 and 3,3'-bis(chlorocarbonyl)azobenzene (Shinkai *et al.*, 1980). Extraction patterns for the alkali metals differed between the (E) and (Z) isomers giving a clear example of photochemical control of the complexation behaviour. Subsequently, the analogue (197) was synthesized in which 2,2'-azopyridine was used for the cap (Shinkai & Manabe, 1984). Photo-

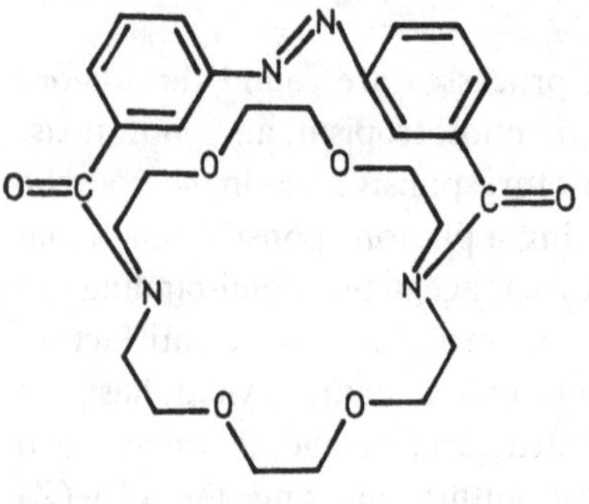

(196) 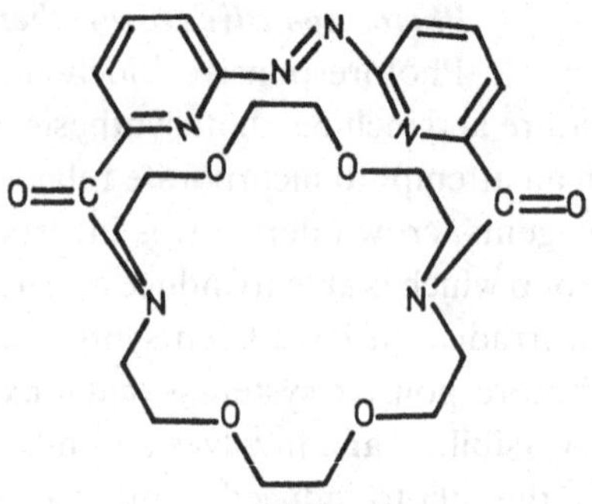   (197)

isomerization of the 2,2'-azopyridine bridge in this molecule leads to a dramatic change in the coordination behaviour of the ligand towards heavy metal ions such as Cu(II), Ni(II), Co(II) and Hg(II). The (E) isomer readily extracts these ions whereas the (Z) isomer barely interacts with them!

For the above system, the affinity for the heavy metals is thus greatly influenced by the configuration of the azo-bridge; in the (Z) form the pyridine nitrogens are presumably no longer correctly orientated for strong coordination to a metal ion held in the cavity of the ligand.

Three 'azobenzeneophane-type' crown ethers in which the 4,4' positions of azobenzene are joined by a polyoxyethylene chain have been synthesized (Shinkai, Minami, Kusano & Manabe, 1983). On irradiation with UV light, the (E) (or *trans*) form (198) is isomerized to the (Z) (or *cis*) isomer (199). The (E) isomer may be regenerated by heating, or by irradiation with visible light: the interconversion is completely reversible.

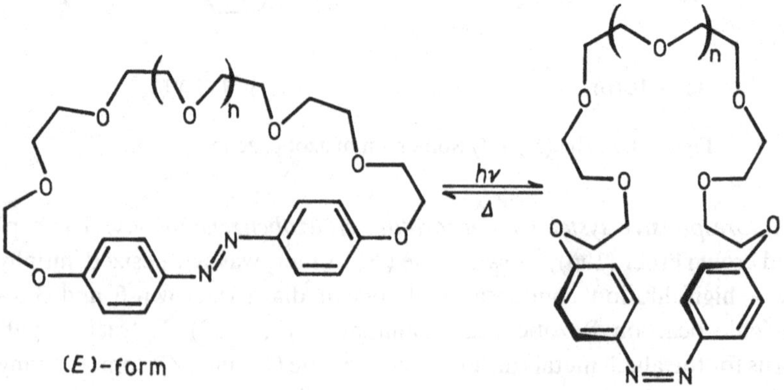

(E)-form

(198)

(Z)-form

(199)

In prior work, related molecules had been obtained in which a polyoxyethylene chain linked the 2,2'-positions of the azobenzene moiety (Shiga, Takagi & Ueno, 1980). However for such systems, the *cis* ⇌ *trans* interconversion is accompanied by gradual photodegradation probably resulting from the presence of steric strain in the *cis* isomers.

A feature of the system illustrated by (198) and (199) is that a large geometrical change may be induced without apparently introducing steric strain into either isomer – the system is the more remarkable because of its 'all or nothing' nature and its high stability towards photodegradation. Thus, solvent extraction experiments indicated that the *trans* isomers show little affinity for alkali metal ions whereas the *cis* forms bind such ions quite well. Molecular models support this finding since it is only for the *cis* forms that the polyether chain may assume a 'loop' configuration suitable for metal-ion complexation. In the (E) isomer the chain is forced into a linearly extended arrangement. As measured by extractability, the (Z) isomers show typical 'hole size' recognition behaviour towards the respective alkali metals. Thus, the isomer with $n = 1$ shows highest affinity for Na^+, with $n = 2$ for K^+ and with $n = 3$ for Rb^+. As might be predicted, the ease of thermally induced (Z) to (E) isomerization is reduced in the presence of alkali ions. This is presumably a consequence of the additional energy required for decomplexation which must necessarily accompany the interconversion. These systems appear to interconvert via an inversion process at one of the nitrogen atoms (rather than by a rotation) such that the polyoxyethylene chain is almost linearly stretched in the inversion transition state.

A range of other derivative systems showing similar behaviour, including ones based on the photoinduced dimerization of anthracene (Desvergne & Bouas-Laurent, 1978), have been investigated. Indeed, the successful demonstration of the general phenomenon has given impetus to further developments in this area. In one study, photochemically induced isomerism has been employed to vary the spatial distance between two crown ether rings attached to the chromophore. In this manner, the ability of the two rings to coordinate to a single metal ion in a 'sandwich' arrangement may be controlled. Several systems of this type based on azobenzene chromophores have been investigated; structure (200) illustrates one example (Shinkai *et al.*, 1982). Collectively, molecules of this general type have been termed 'butterfly' crown ethers. For such molecules, complexation is once again dependent on the radius of the alkali metal involved and on the ease with which interconversion from (E) to (Z) forms occurs; for 'sandwich' coordination of (200) the (Z) configuration must be present – see (201).

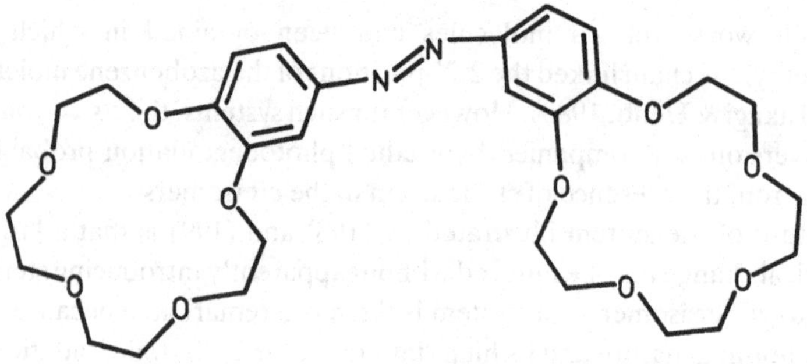

(200)

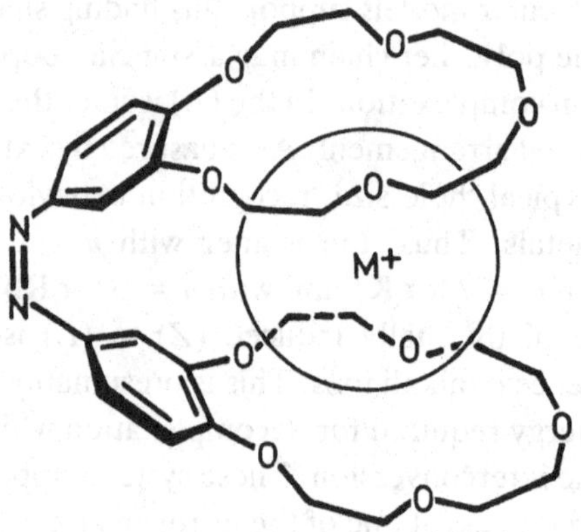

(201)

Light-driven membrane transport. Cations may be transported through liquid membranes using crown ethers. For example, a typical system is of the type water-phase(I)/organic-phase/water-phase(II). The metal ion is added to water-phase(I) and the crown ether to the organic phase (to yield the liquid membrane). The crown acts as carrier for metal ions from water-phase(I) across the liquid membrane phase into water-phase(II). There have now been a very large number of studies of this type reported and a fuller discussion of this topic is given in Chapter 9.

Because the metal ion has to be *taken up* at the water-phase(I)/organic-phase interface but then *released* at the organic-phase/water-phase(II) interface, the best carrier for ion transport is usually a crown giving only a moderately stable complex with the ion (not one of high stability). For systems of this type, plots of log K for metal-crown binding versus ion-transport rate for different crowns normally reach a maximum for moderate K values (Kirch & Lehn, 1975; Kobuke *et al.*, 1976). Thus in designing a carrier macrocycle there is a dilemma – strong metal-ion binding will favour the ion 'take-up' step but not favour the 'ion-release' step. Several studies have attempted to overcome this dilemma by incorporating on–off switches of the type just discussed in the carrier molecule (Shinkai & Manabe, 1984). That is, light is used to control metal-ion binding such that strong binding occurs for metal uptake but weaker binding is induced for the metal-release step. Various systems of this type have now been reported: in some, UV light is used to induce photo-isomerization in one direction while visible radiation results in back-isomerization. In other systems, UV light/thermal isomerization is employed while in others pH changes are also superimposed.

An example of light-assisted transport of the first type involves (200) as the carrier in the liquid membrane. In this case, irradiation of the membrane alternatively with UV and visible light significantly increases the rate of K^+ and Rb^+ transport in the presence of picrate ion. This system also exhibits discrimination since the transport of K^+ is favoured over Rb^+ (Shinkai, Shigematsu, Sato & Manabe, 1982).

Binucleating crown ethers

The major categories. As for the other category of macrocycles incorporating mainly N, S, P, or As donors, a number of binucleating crown systems are also known. As discussed in Chapter 3, such binucleating systems may be considered to fall (mainly) into three broad categories: (i) large-ring crowns which incorporate two metal ions in the macrocyclic ring, (ii) two-ring systems linked by a bridge moiety, and (iii) two-ring systems linked by more than one bridge often forming a 'face-to-face' configuration.

A few mixed-donor binucleating macrocycles incorporating some ether oxygen functions have already been mentioned in Chapter 3. This earlier discussion is now expanded to include cyclic systems in which the donor sets consist (totally or predominantly) of ether oxygens.

A large ring system incorporating two metal ions. An example of the large-ring category mentioned above is given by the 2:1 complex of potassium thiocyanate with dibenzo-24-crown-8. In the solid state, each K^+ ion is contained in the macrocyclic ring of this crown. Both thiocyanate ions and two of the ether donor atoms form bridges between K^+ ions such that each achieves a coordination sphere of five oxygens in a plane with thiocyanate nitrogens above and below this plane (Fenton, Mercer, Poonia & Truter, 1972). The structure is illustrated in Figure 4.8.

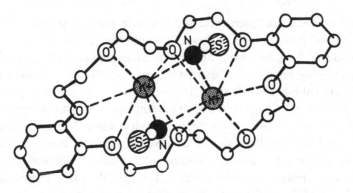

Figure 4.8. The 2:1 complex of KNCS with dibenzo-24-crown-8 (Fenton, Mercer, Poonia & Truter, 1972).

Dual ring systems incorporating a single bridging moiety. An example from the second category of binucleating macrocycles (those incorporating dual rings) is given by (202) (Handyside, Lockhart, McDonnell & Rao, 1982). This ligand is one of a series of such molecules containing different polyether ring sizes as well as different numbers of methylene groups joining the rings. These bis-macrocyclic species were synthesized in order to promote the formation of 'clam' (or 'ear muff') type geometries with particular alkali ions. In such an arrangement, the metal ion is held between the two macrocyclic rings in a sandwich configuration (with the respective rings connected by the polymethylene bridge). Physical measurements indicate that, when the cation is of the 'correct' size relative to the dimensions of the bis-crown, then 'clam' coordination behaviour does occur. Solvent extraction studies indicated that these compounds bind to the alkali metal ions; however, those ions which are large enough to form 'clam' arrangements are favoured relative to those which are not. The formation or non-formation of a 'clam' structure thus serves as a discrimination mechanism for particular alkali ions.

(202)

(203)

The cyclohexane-containing system (203) was also prepared in an attempt to obtain metal complexes with 'clam' type geometries (Owen, 1983). As for the previous system, it was considered that such complexes might show enhanced shielding of the cation, from both the solvent and the counter ion present, but still allow the bound metal to be readily released on demand. As is evident from our earlier discussion, both these are desirable properties for metal-ion transport systems.

For (203), models indicated that the isomer containing *cis-syn-cis* hydrogen atoms on the cyclohexane ring should be able to form clam-type complexes, provided the cyclohexane ring is in the flexible or 'twist' conformation. The models suggested that the cavity defined by the ten oxygen donors would be ideal for K^+. However, for the potassium and barium thiocyanate complexes, configurations of type (204) do not occur in the solid state. Instead, two molecules of the bis-crown coordinate simultaneously to two alkali metal ions – both these 2:2 complexes have structures of type (205).

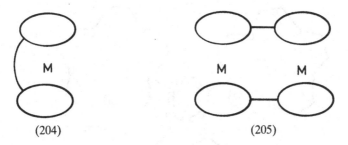

(204)   (205)

The spiro-derivative (206) forms anhydrous complexes with larger alkali ions (such as K^+). Such complexes appear to have a 2:1 structure of type (207) (Weber, 1979). For Li^+, the hydrated complex $[Li_2L(H_2O)_4]I_2$ [where L = (206)] is formed. The X-ray structure of this species reveals that it is essentially of type (207) except that the Li^+ in each macro-ring coordinates to only three ether oxygens of the ring. Each Li^+

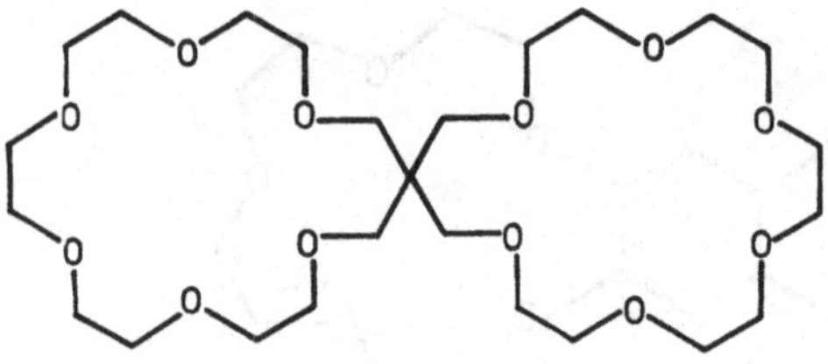

(206)

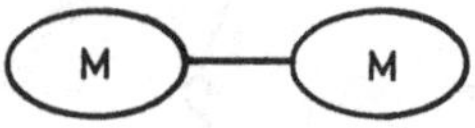

(207)

also binds to two water molecules, one of which is hydrogen-bound, via both its hydrogens, to two further ether groups in the ring. Overall, each lithium is five-coordinate with a distorted trigonal bipyramidal structure (Czugler & Weber, 1981).

Polynucleating crowns

Crown derivatives which are able to bind simultaneously to more than two metal ions have been synthesized; such species have been termed 'multiloop crowns' (Weber, 1982). Polynuclear cation receptors of this type provide yet another series of novel crown derivatives whose metal complexes tend to exhibit a range of unusual properties. Because of the diversity of the structures falling in this category, only a brief mention of this ligand type is presented here.

Related structures to the binucleating crown (206), in which three or four individual macrocycles are linked by spiro-groups, have been synthesized (Weber, 1979; Weber, 1982). Other ligands of this general type, showing a variety of topological arrangements, are also known (Weber &

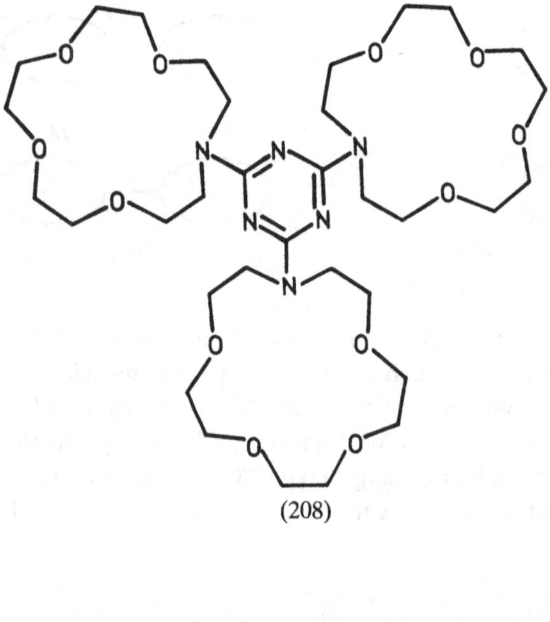

(208)

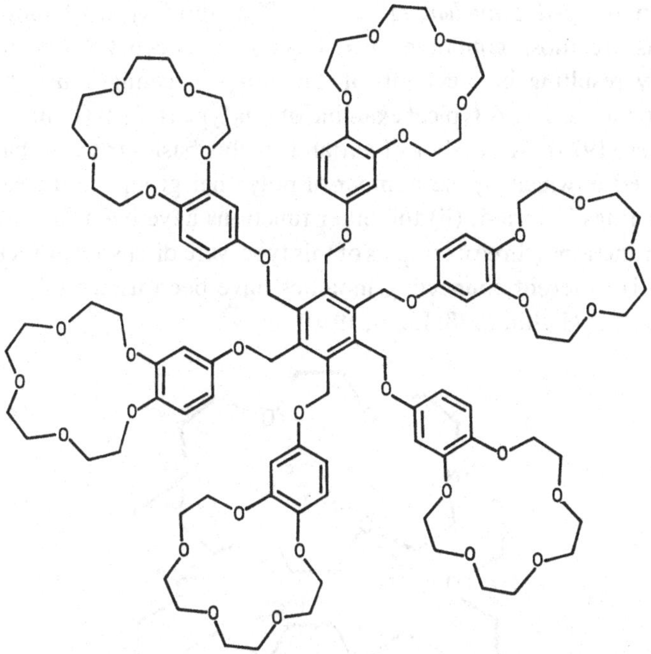
O

(209)

Vogtle, 1981; Davies, Kemula, Powell & Smith, 1983). Two such ligands are (208) and (209).

Further elaborate molecules of the present category continue to be synthesized. For example, the 'Mobius strip' crowns (210) and (211) were obtained using the procedure outlined by [4.6] (Walba, Richards & Haltiwanger, 1982).

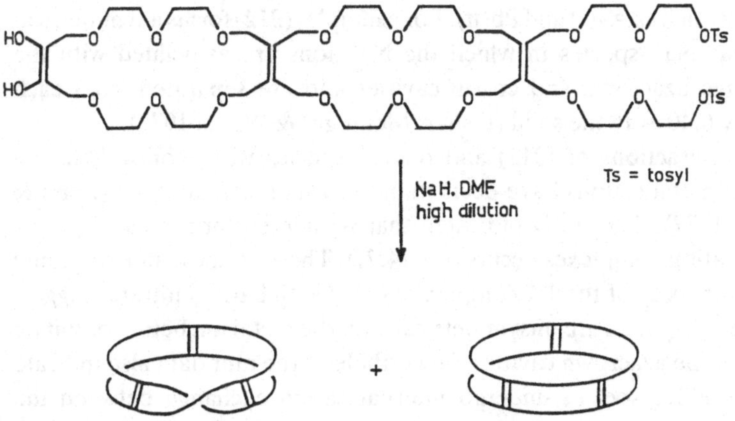
HO
HO
OTs
OTs
Ts = tosyl
NaH, DMF
high dilution
+

[4.6]

(210) (211)

Doubly-bridged binucleating crowns. The third type of binucleating ligands are those containing two linkages between the crown rings – usually resulting in a cylindrical 'face-to-face' configuration for these bis-crown species. A typical example of this type is (212) (Lehn, Simon & Wagner, 1973). A number of variants of this basic structure have been prepared in which (i) the number of polyether groups in the respective crown rings is varied, (ii) thioether functions have been substituted for ether functions (three examples of this type were discussed in Section 3.4) and (iii) different connecting moieties have been used to link the two crown rings (Lehn, 1978; Lehn, 1980).

(212)

Systems of the above type incorporating larger rings form binuclear complexes with a number of alkali and alkaline earth metal ions as well as with ions such as Ag(I) and Pb(II). For example, (212) forms a symmetrical dinuclear Na$^+$ species in which the Na$^+$ ions are associated with the respective azacrown and crown cavities and are separated from each other by 6.40 Å in the solid (Fisher, Mellinger & Weiss, 1977).

The interactions of (212) and related species with monovalent and divalent metal cations have been studied by nmr spectroscopy (Lehn & Simon, 1977). The study indicated that sequential formation of mono- and di-cation complexes occurs (see [4.7]). These studies, and especially a ^{13}C nmr study of the 1:1 complex of (212) with barium nitrate, suggest that the 1:1 species are unsymmetrical with the metal ion being contained in one of the azacrown cavities. Nevertheless, the nmr data also indicate that these 1:1 species undergo internal cation exchange between the respective azacrown sites. This intramolecular dynamic behaviour serves

[4.7]

as a model for the elementary 'jump' processes which occur for metal ions between binding sites in channel-type membrane systems (Lehn, 1978) – see Chapter 9.

Heteronuclear species may also be prepared – the nmr studies indicate that a heteronuclear Ag(I)/Pb(II) species is produced in solution when a 1:1 mixture of the respective dinuclear Ag(I) and Pb(II) complexes is dissolved. However, the reaction is not quantitative and the heteronuclear complex exists in equilibrium with both of its homonuclear precursors.

Not surprisingly, the stabilities of the 1:1 complexes of (212) with the alkali and alkaline earth metals parallel those for N,N-dimethyl-18-diazacrown-6. Further, the addition of a second metal ion to yield the corresponding dimetallic, homonuclear species occurs readiy (Lehn & Stubbs, 1974). Indeed, the strength of binding of the second cation (K_2 value) tends to be little different from that of the first. Apparently, the spacing between the azacrown moieties [in (212) and similar derivatives] is large enough to render insignificant the electrostatic (repulsive) effects which will occur between the two bound cations.

4.2   The cryptands
The diaza-cryptands
Following the work of Pedersen involving two-dimensional crowns, Lehn and coworkers developed a range of three-dimensional,

polycyclic ligand systems which they named **cryptands** (Greek: *cryptos* = cave). The metal complexes of these cage ligands are called **cryptates** (Dietrich, Lehn & Sauvage, 1969). The development of such systems was of major significance to the field of polyether metal-ion chemistry and, subsequently, an extremely large number of studies involving such species have been carried out.

The 'parent' series of cryptands may be represented by (213) – a stepwise increase in cavity size occurs along this series. High-dilution procedures are employed for the synthesis of these cages.

m	n	Abbreviation	Estimated cavity diameter (Å)
0	0	1.1.1	1·0
0	1	2.1.1	1·6
1	0	2.2.1	2·2
1	1	2.2.2	2·8
1	2	3.2.2	3·6
2	1	3.3.2	4·2
2	2	3.3.3	4·8

(213)

A typical synthesis. The preparation of (213; $m, n = 1$) serves to illustrate the general synthetic strategy used to obtain this class of polyether cage. Starting from diaza-18-crown-6 and the required diacid chloride, condensation followed by reduction yields the corresponding cryptand, 2.2.2 (Dietrich, Lehn & Sauvage, 1970; Dietrich, Lehn, Sauvage & Blanzat, 1973) – see [4.8].

Cryptands of this type are able to exist in three isomeric forms since each of the bridgehead nitrogens may be orientated inwards or outwards with respect to the molecular cavity – the three isomers are thus 'in-in', 'in-out', and 'out-out'. In the solid state, 2.2.2 has been shown to have an 'in-in' arrangement (Metz, Moras & Weiss, 1976).

Protonation of cryptands. Novel protonation behaviour has been observed for these (diaza) cryptands (Cheney & Lehn, 1972; Kresge, 1975). Internal protonation of ligands such as 1.1.1 and 2.1.1 leads to species exhibiting very slow proton exchange rates. In contrast, for simple systems, proton exchange rates are invariably very fast. For 1.1.1, the

[4.8]

deprotonation of the doubly protonated species remains very slow even in the presence of strong base. This reluctance to lose protons is a reflection of the 'tight' cage formed by this system in its 'in-in' configuration.

Metal complex formation. The cryptands readily form complexes with a range of metal ions, provided the ion involved is not too large to be contained in the macrocyclic cavity. Complexation of the alkali and alkaline earth metals may be readily followed by monitoring the changes in the ^1H and ^{13}C nmr spectra of the cryptands in the absence and presence of these ions (Dietrich, Lehn & Sauvage, 1969; Dietrich, Lehn & Sauvage, 1973a). A range of crystal structure determinations confirm that the metal ion normally resides in the central cavity of such ligands. The X-ray structure of the Rb^+ complex of 1.1.1 is shown in Figure 4.9 (Moras, Metz & Weiss, 1973); note that, as usual, the ligand is coordinated as its 'in-in' isomer in this structure.

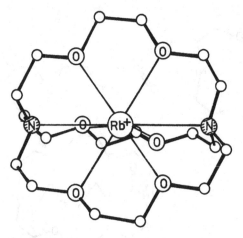

Figure 4.9. Structure of the Rb^+ complex of cryptand 1.1.1 (Moras, Metz & Weiss, 1973).

Complexes of the cryptands having 2:1 stoichiometries are also known; for example, with $Pb(\text{II})$, 2.1.1 forms a species of type $[Pb_2(2.1.1)]^{4+}$ in which both $Pb(\text{II})$ ions appear to lie outside the macrocyclic cavity (Arnaud-Neu, Spiess & Schwing-Weill, 1982).

The stability of cryptate complexes. The cage topology of the cryptands results in them yielding complexes with considerably enhanced stabilities relative to the corresponding crown species. Thus the K^+ complex of 2.2.2 is 10^5 times more stable than the complex of the corresponding diaza-crown derivative – such enhancement has been designated by Lehn to reflect the operation of the **'cryptate effect'**; this effect may be considered to be a special case of the **'macrocyclic effect'** mentioned previously.

The range of cryptands given by (213) show strong selectivity towards individual alkali and alkaline earth metal ions. Thus, the smaller systems provide classic examples of alkali-metal selectivity based on cavity size. As the cavity size increases along the series 2.1.1, 2.2.1, and 2.2.2, preferential complexation of Li^+, Na^+ and K^+ respectively, occurs (Lehn & Sauvage, 1975). Relative to the crowns, the macrocyclic cavity is more clearly defined in the cryptates and, hence, 'best fit' behaviour tends to occur more readily. Cryptate complexation involving a range of transition and other heavy metal ions has been investigated and, in a number of cases, complexes of moderate stability are formed (Arnaud-Neu, Spiess & Schwing-Weill, 1977; Anderegg, 1981). Similarly, the cryptands also form stable complexes with the lanthanide ions (Almasio, Arnaud-Neu & Schwing-Weill, 1983; Burns & Baes, 1981; Yee, Gansow & Weaver,

1980). Further discussion of the stability of cryptand complexes is presented in Chapter 6.

Solubilization and phase transfer catalysis. As mentioned in earlier parts of this chapter, the cryptands, like the crowns, are also effective in solubilizing a range of inorganic salts in both polar and non-polar solvents. For instance, in the presence of 2.2.2, the solubility of $BaSO_4$ in water is increased by approximately 10^4 (Lehn, 1978). Similarly, as with certain crowns, $KMnO_4$ may be taken up into solvents such as benzene in the presence of 2.2.2. Since the cryptand totally encloses the complexed cation, efficient shielding of the ion occurs. This is well illustrated by the behaviour of the ^7Li nmr signal for the Li^+ complex of 2.1.1 in different solvents: the chemical shift of this signal is quite independent of the nature of the solvent used (Cahen, Dye & Popov, 1975).

For those applications involving the activation of an inorganic anion (that is, generation of a 'naked' anion), the cryptands, rather than the crowns, tend to be the reagents of choice. Such reagents are thus also ideal for applications involving phase-transfer catalysis of the type discussed previously.

Other cryptand derivatives. Substituents on the cryptand structure affect, in a variable manner, the stabilities of the corresponding complexes. Thus, fusing benzo rings to the polyether bridges in cryptands is expected to decrease the macrocyclic cavity size slightly and also to reduce the donating ability of the adjacent ether oxygens (reflecting the electron-withdrawing nature of the aromatic rings). Hence, it is not surprising that incorporation of a benzo group in 2.2.2 to yield (214), alters the selectivity of this ligand towards the alkali metals. However, it is more difficult to assess the relative importance of the above-mentioned contributions for a particular metal-ion system. For (214), the Na^+ complex is found to be more stable (in methanol) than is the complex of this metal with 2.2.2. In contrast, the complexes of (214) with K^+ and Ba^{2+} are both less stable than those of 2.2.2 (Dietrich, Lehn & Sauvage, 1973b).

A variety of other cryptand derivatives have been reported. Two systems incorporating carbon bridgeheads are (215) (Coxon & Stoddart, 1977) and (216) (Parsons, 1978; Herbert & Truter, 1980). Both ligand types form complexes with metals such as Na^+ and K^+. Other derivatives containing additional nitrogen donors are (217) (Newkome *et al.*, 1979), (218) (Buhleier, Wehner & Vogtle, 1978) and (219) (Lehn & Montavon, 1978). As for the simple crowns, substitution of nitrogen donors for ether oxygens results in systems which exhibit weaker binding towards the

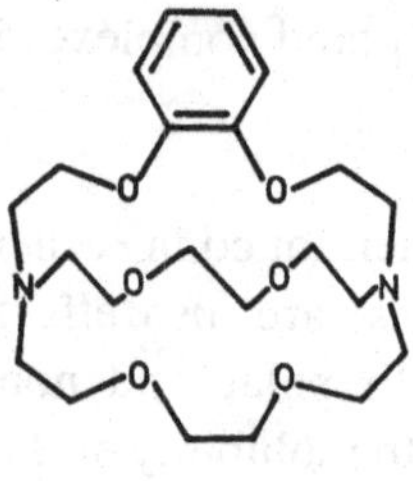

(214)

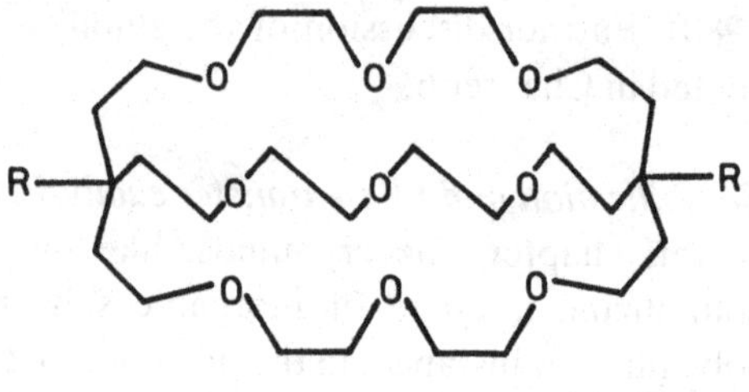

(215)

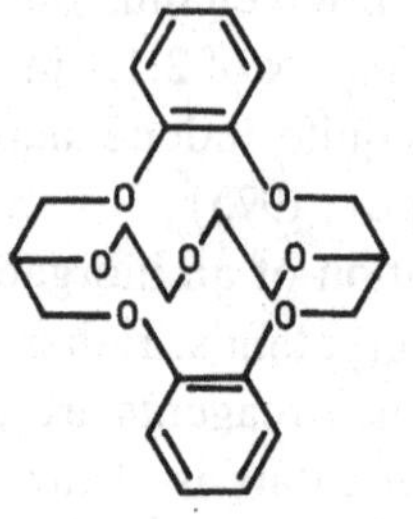

(216)

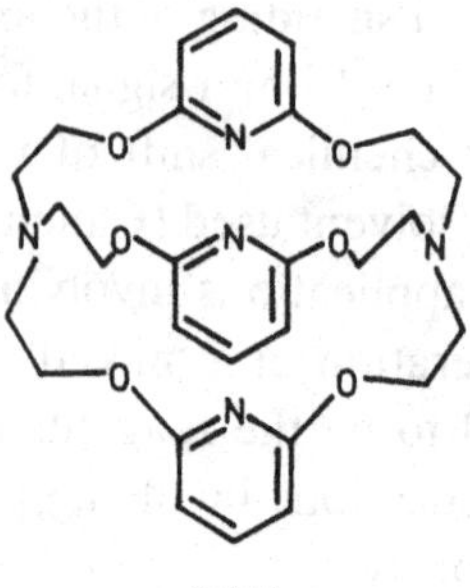

(217)

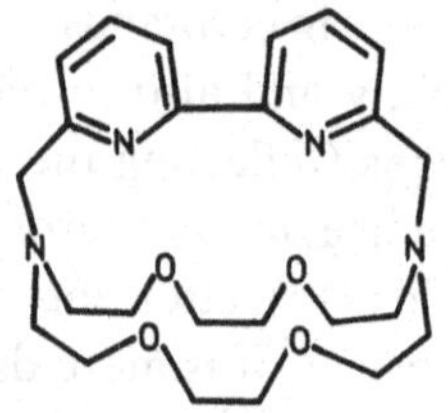

(218)

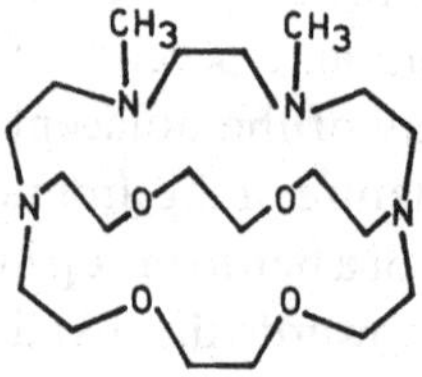

(219)

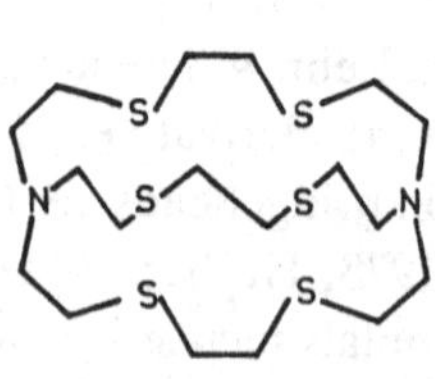

(220)

alkali and alkaline earth ions. A similar consideration applies when thioether donors are substituted for ether donors as, for instance, has occurred in the N_2S_6-system (220) (Lehn, 1978).

Cryptands of the type (217)–(220) tend to form stable complexes with a number of heavy metal ions. Of particular interest is the selectivity of (219) for Cd(II); the complex of this metal is approximately 10^6–10^7 times more stable than its complexes with either Zn(II) or Ca(II). This reagent may prove useful for removing toxic Cd(II) from biological systems as well as for other applications involving sequestration of this ion (for example, in antipollution systems). The selectivity observed in the above case appears to arise because: (i) the nitrogen sites favour coordination to Zn(II) and Cd(II) relative to Ca(II) and (ii) the cavity size favours coordination of Cd(II) relative to Zn(II).

Immobilized cryptates. Like the crowns, cryptates have been immobilized on polymeric backbones. A typical system is given by (221) (Cinquini, Colonna, Molinari, Montanari & Tundo, 1976). In this case, the polymeric matrix is polystyrene cross-linked with *p*-divinyl benzene and the cage is connected to this matrix via a long-chain aliphatic spacer group. This reagent is quite effective as a (triphase) transfer catalyst.

(221)

A number of other cryptand-bound polymers have been synthesized using similar procedures to those discussed previously for immobilization of crown molecules. Apart from their use in phase transfer catalysis, such polymers have been studied extensively as chromatography reagents for the separation of a range of metal-ion types (Blasius & Janzen, 1982); in a number of instances quite useful separations have been achieved.

4.3   Generation of alkalides and electrides

Novel anions stabilized by alkali-polyether cations

The ability of a crown (such as 18-crown-6) or a cryptand (such as 2.2.2) to shield an alkali cation by complex formation has enabled the synthesis of a range of novel compounds containing an alkali metal cation coordinated to a crown or cryptand for which the 'anion' is either a negatively charged alkali metal ion or a single electron (Dye & Ellaboudy, 1984; Dye, 1984). Such unusual compounds are called 'alkalides' and 'electrides', respectively.

Background alkali metal chemistry. The alkali metals have the lowest ionization potentials of any group in the periodic table and hence their chemistry is dominated by the M^+ oxidation state. However, it has been known for some time that a solution of an alkali metal (except lithium) in an amine or ether forms not only M^+ ions and solvated electrons but also alkali anions of type M^- (Matalon, Golden & Ottolenghi, 1969; Lok, Tehan & Dye, 1972). That is, although an alkali metal atom very readily loses its single s-shell electron:

$$M \rightarrow M^+ + e^-$$

it is nevertheless also able to gain an electron (to yield a filled outer s-shell):

$$M + e^- \rightarrow M^-.$$

Even though alkali metal anions may be freely generated in solution, the isolation of solid salts containing such anions is not straightforward. Thus, the disproportionation observed when an alkali metal is dissolved in an amine or ether

$$2M_{(s)} + n(\text{solvent}) \rightleftharpoons [M(\text{solvent})_n]^+ + M^-$$

appears to be largely controlled by favourable solvation of the M^+ cation. Removal of the solvent in attempting to isolate a salt of M^- simply reverses the above equilibrium and the free metal tends to be recovered instead.

Polyether complexation. The solution of the above problem is to add a suitable crown ether or cryptand to the alkali metal solution. This results in complexation of the alkali cation and apparently engenders sufficient stabilization of the M^+ cation for alkalide salts of type $M^+L.M^-$ (L = crown or cryptand) to form as solids. Thus the existence of such compounds appears to reflect, in part, the ability of the polyether ligands to isolate the positively charged cation from the remainder of the ion pair.

The alkalides. The first crystalline alkalide to be prepared in this manner was [Na⁺(2.2.2)].Na⁻. This salt is obtained as shiny, gold-coloured crystals (Dye *et al.*, 1974). The ^{23}Na nmr spectrum yields a narrow upfield signal for the Na⁻ ion (Dye, Andrews & Ceraso, 1975); the X-ray structure indicates close-packed sodium cryptate cations with Na⁻ anions occupying octahedral holes between the cryptate layers (Tehan, Barnett & Dye, 1974).

Although crystalline potassides, rubidides and caesides have all been prepared, sodides are more readily isolated since they tend to be the more kinetically and thermodynamically stable. A number of mixed-metal alkalides has also been isolated but they are all of type $M^+L_nNa^-$ ($n = 1$ or 2).

The electrides. Following many experimental difficulties, Dye and co-workers were able to demonstrate that reaction of 18-crown-6 with caesium in a 1:2 ratio (under specified conditions) leads to isolation of shiny, black crystals of a product of composition $Cs^+(18\text{-crown-}6)_2$ (El-laboudy, Dye & Smith, 1983; Dye & Ellaboudy, 1984). The solid-state ^{133}Cs nmr spectrum, the esr spectrum, and the magnetic susceptibility of this product all indicated that it was of type $Cs^+(18\text{-crown-}6).e^-$; that is, a crystalline electride in which the 'anion' is a single electron. In overall terms, this exotic class of compound may be considered to lie on the border between metals and non-metals.

4.4   Concluding remarks

This chapter has largely been concerned with two- and three-dimensional polyether ligand systems and their complexes with a range of metal ions. A variety of other related ligand systems are also known but in many instances these have been primarily used to form complexes with non-metallic guests.

The formation and properties of a range of host-guest complexes involving macrocyclic hosts and non-metallic guests are described in the next chapter.

5

Host-guest chemistry: macrocyclic hosts and non-metallic guests

5.1   Introductory remarks

Host-guest interactions. The area of host-guest chemistry encompasses the complexation by organic hosts of a range of both organic and inorganic guests.

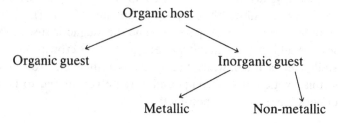

The interaction of macrocyclic hosts with metal-ion guests was discussed in Chapter 4 while, in this chapter, the discussion is extended to include guests other than metal ions. The complexation of non-metallic guests by organic hosts has received much attention in recent years. A variety of interaction types may contribute to such host-guest binding. Thus, as well as ion-pairing and hydrogen bond formation, such interactions may involve π-acid and π-base attractions, as well as van der Waals attractive forces, and may also be aided by favourable solvation/desolvation effects. With respect to the latter, a hydrophobic guest may interact with a hydrophobic host and such 'hydrophobic' interactions may be a major driving force for molecular association. Nevertheless, they appear to reflect the fact that the molecular complex perturbs the solvent less than do the separated host and guest. Because of their rather non-specific nature, such hydrophobic effects are not expected to engender much selectivity; however, they will usually contribute substantially to the overall strength of a host-guest interaction.

For selective hosts, the receptor cavity will incorporate sensing as well as binding elements (these may or may not be combined) such that potential guests are rejected or accepted on the basis of their spatial and electronic properties. That is, for strong host-guest complexation, the host and guest must be capable of interacting with each other such that the respective binding sites exhibit complementary steric and/or electronic arrangements.

Relationship to biochemical systems. Host-guest chemistry constitutes a field of endeavour in which many of the examples display something of the elegance and subtlety characteristic of the interaction of substrates with biological receptor sites. Thus many host-guest interactions of the above type, especially those involving organic guests, may provide simple models for the often more complex interactions which are a feature of particular biochemical processes (Stoddart, 1979; Cram & Trueblood, 1981; Meade & Busch, 1985; Sutherland, 1986). Indeed, molecular complex formation plays a central role in such important areas as the mechanism of drug action, enzyme-substrate interactions, immunological responses, and the storage and retrieval of genetic information. All of these are characteristically associated with strong discriminatory behaviour – a feature which is also shown by many of the synthetic systems.

Other considerations. For particular synthetic complexes, the reactivity of the guest has been investigated when complexed and, in some instances, a considerable rate increase is observed relative to that for the uncomplexed guest. When this occurs, it may reflect that the design of the host leads to stabilization of the rate-limiting transition state. In these cases the systems can be thought of as exhibiting rudimentary 'enzymic' catalytic behaviour. Indeed, the prospect of designing systems which promote enzyme-like rate enhancements for simple chemical processes has provided a strong motivation for a range of such studies. This area shows especial potential since it has strong overtures for the development of new chemical processes which are both more specific and also of lower energy requirements than presently existing ones. Of course, besides catalysis, there is a wide range of potential medical, analytical and other industrial applications which exist for particular host-guest systems of the present type.

5.2   Host–guest complexation involving crowns and related polyether ligands

Crown polyether hosts

Apart from complex formation involving metal ions (as discussed in Chapter 4), crown ethers have been shown to associate with a variety of both charged and uncharged guest molecules. Typical guests include ammonium salts, the guanidinium ion, diazonium salts, water, alcohols, amines, molecular halogens, substituted hydrazines, *p*-toluene sulfonic acid, phenols, thiols and nitriles.

With simple crowns, complex formation may involve various degrees of inclusion of the guest into the cavity of the crown and a conformational rearrangement of the crown is almost always necessary for strong complexation to occur. This will normally involve a redirection of the donor electron pairs on complex formation so that their final orientations optimize a particular host-guest interaction.

Complexes of alkylammonium and other cationic guests. Complexes of protonated amines with crown ethers have been studied since the beginning of crown chemistry (Pedersen, 1967). For such adducts, the presence of hydrogen bonding between the protons of the substrate and some of the ether oxygens of the macrocycle almost certainly contributes markedly to their stability (Cram & Cram, 1978). For example, addition of the methylammonium cation to 18-crown-6 yields a 1:1 adduct. From inspection of CPK (Corey–Pauling–Koltun) space-filling molecular models, it was suggested that the binding between $CH_3NH_3^+$ and 18-crown-6 involves hydrogen bond formation between the three ammonium hydrogens and alternate oxygens in the macrocyclic ring, as shown by (222). The host molecule will be approximately planar with the nitrogen of the guest situated slightly above this plane at the apex of a shallow pyramid. Since the nitrogen and its attached methyl group protrudes above one face of the host, a complex of this type has been termed a **perching**

(222)

complex by Cram (as opposed to a **nesting** complex which will have over half of the guest surface in contact with the host). A similar arrangement has been confirmed by X-ray diffraction to occur in a related complex containing NH_4^+ as the complexed guest (Nagano, Kobayashi & Sasaki, 1978) as well as in the complex between 2,3-naphtho-18-crown-6 and $CH_3NH_3^+$. The ^1H nmr spectrum of this latter species in $CDCl_3$ confirms that the solid-state structure persists in solution (Cram & Trueblood, 1981). Although hydrogen bonding appears to be the dominant cause of the considerable stability of all these adducts, a further contribution may involve electrostatic attraction between the positively charged guest and the negative dipoles associated with the electronegative oxygens of the ring (Timko *et al.*, 1977).

Binding constants corresponding to:

$$\text{host} + \text{guest} \overset{K}{\rightleftharpoons} \text{molecular complex}$$

for the interaction of a number of such crown-hosts and alkylammonium guests have been determined. Such studies have confirmed that the molecular match of the host for the receptor sites of the guest is an important factor in the formation of a strong inclusion complex. Thus 18-crown-6 has been demonstrated (Timko *et al.*, 1977) to bind t-$C_4H_9NH_3^+$ more than 10^4 times more strongly than does its open-chain analogue (223). This difference apparently reflects the presence in the former of an existing cavity which results in less rearrangement being necessary on binding to the ammonium cation. Indeed, open-chain ligands such as (223) rarely yield thermodynamically stable host-guest complexes when organic hosts are involved.

(223)

A large number of other host compounds have been designed to incorporate functional sites which are suitably placed for interaction with the binding sites of a guest. Molecular models suggested (Newcomb, Timko, Walba & Cram, 1977) that the somewhat more rigid ring in the

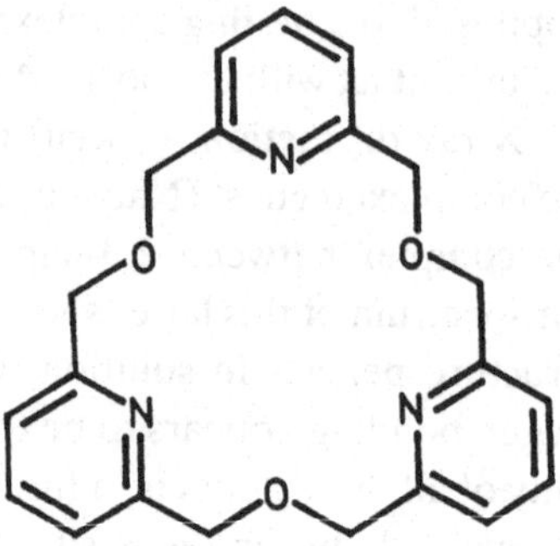

(224)

aza-crown (224) would provide either three oxygens or three nitrogens in ideal positions for hydrogen bond formation with the three pyramidal ammonium protons of a RNH_3^+ species. Indeed, this macrocycle yields quite strong 1:1 complexes. Of the two potential modes of bonding, the latter, with NH$\cdots$N hydrogen bonds, is probably the more stable.

The double-ring host (225), incorporating a chiral 1,1-dinaphthyl (S form) backbone, is capable of complex formation with a difunctional alkylammonium ion of type $H_3^+N–R–NH_3^+$ [where R is $-(CH_2)_n-$ ($n = 4$ or 5) or m–C_6H_4 or p–C_6H_4]. These complexes contain the guest aligned in a perching arrangement between the two crown 'jaws' as shown in (226)

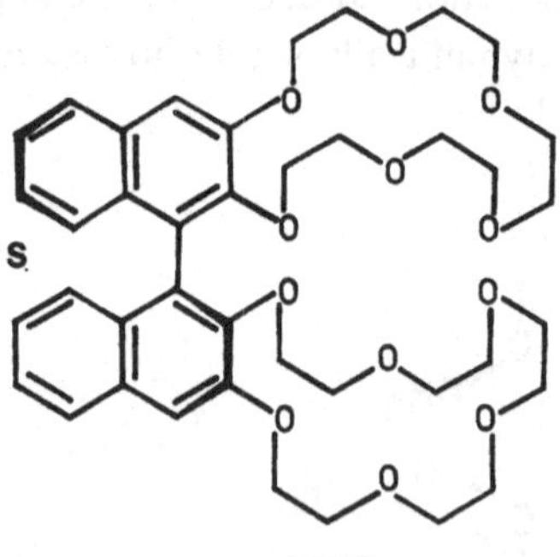

(225)

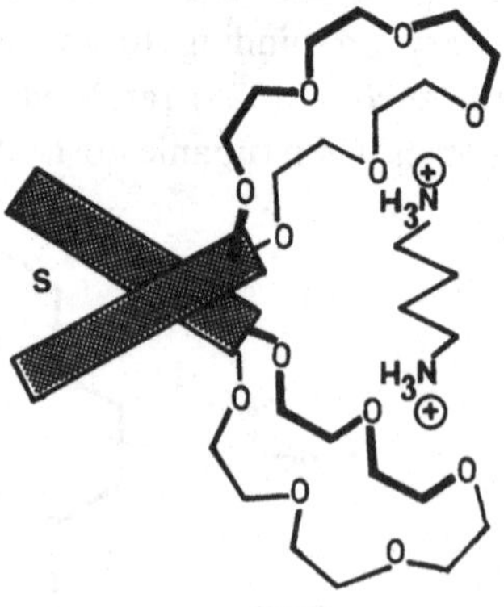

(226)

(Helgeson, Tarnowski & Cram, 1979). X-ray diffraction confirms that a similar structure occurs in the solid state: in this instance, the tetra-methylene diammonium ion is the guest (the dipositive charge on the complex is balanced by a pair of uncoordinated hexafluorophosphate counter ions) (Goldberg, 1977).

A further class of host molecules contain an ionizable functional group

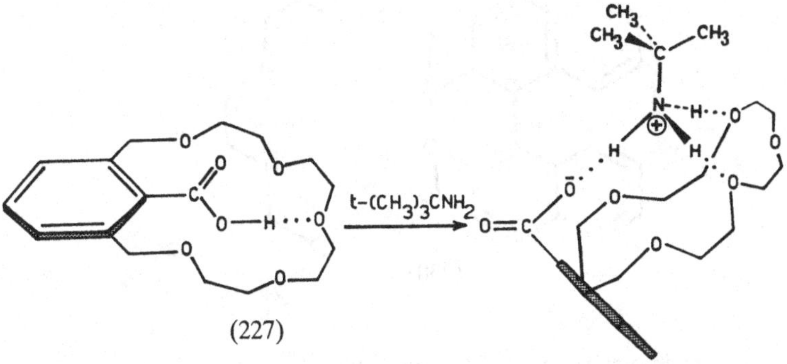

(227)

(228)

[5.1]

in their structures. Thus, when complexation involves a singly-charged cationic guest, there will be no need for a separate counter ion. A well-studied example of this type involves the carboxylate-containing host (227) (Newcomb, Moore & Cram, 1977). This species, in its deproto-nated form, readily forms 1:1 complexes with alkylammonium cations such as $(CH_3)_3CNH_3^+$ [5.1]. For the complex of this latter guest, the perching structure (228) occurs (Goldberg, 1975a).

The X-ray structure of the molecular complex between the hydrazinium cation (as its perchlorate salt) and 18-crown-6 also reveals a beautiful complementary relationship between the crown and this guest. The $H_2NNH_3^+$ ion is contained in the cavity of the crown such that all five hydrogens of the hydrazinium ion are bound to five different oxygens of the crown. The sixth crown oxygen is somewhat further away (3.33 Å) from its nearest nitrogen than are the two oxygens hydrogen-bound to the $-NH_2$ moiety (for which the N–H$\cdots$O distance is 3.05 Å in each case). Structure (229) is a diagramatic representation of the bonding in this complex (Trueblood, Knobler, Lawrence & Stevens, 1982).

From models, it is apparent that 18-crown-6, with all oxygen electron pairs pointing inwards, also provides an ideal cavity for the diazonium

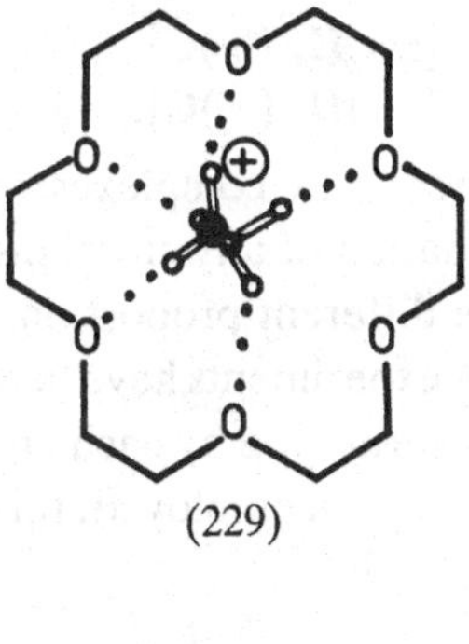

(229)

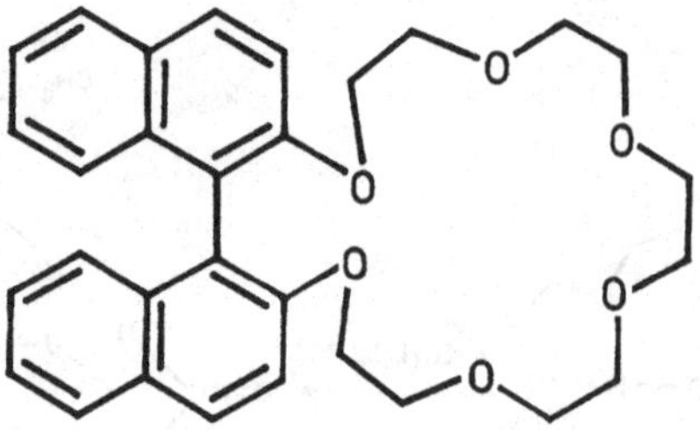

(230)

function of $C_6H_5N\equiv N^+$ (Kyba *et al.*, 1977). Thus, a solution of this crown in chloroform solubilizes otherwise insoluble *p*-toluenediazonium tetrafluoroborate. Similarly, host (230) also solubilizes $(C_6H_5N\equiv N)BF_4$. The need for a complementary cavity is strikingly illustrated by the observation that the crown analogue containing one less $-CH_2CH_2O-$ group in the ring does not solubilize this salt; the hole is now too small to embrace the $-N\equiv N^+$ group which can be considered to be a cylinder of about 2.4 Å diameter. The open-chain compound (231) also fails to solubilize $(C_6H_5N\equiv N)BF_4$.

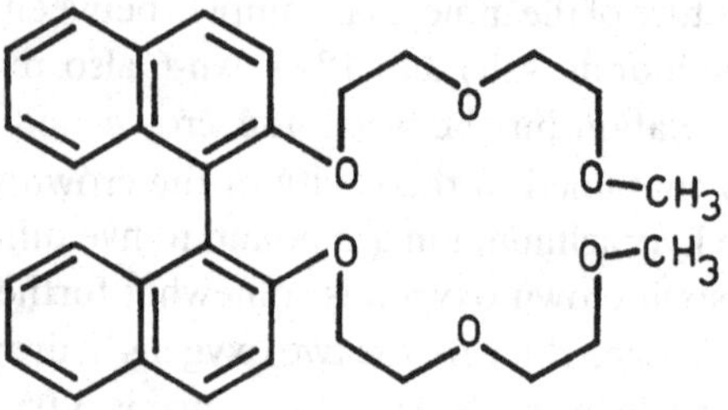

(231)

Systems exhibiting chiral recognition. Complexation of an optically active guest molecule $(+)G$ or $(-)G$ by a chiral host $(-)H$ may be represented as follows:

$$(-)H + (+)G \rightleftharpoons [(-)H, (+)G]$$
$$(-)H + (-)G \rightleftharpoons [(-)H, (-)G].$$

In theory, the two diastereomeric complexes will have different association constants. The evaluation of any chiral discrimination will depend upon measurement of the different proportions of the diastereoisomers formed. For example, nmr experiments have been successful in determining the degree of complex formation by each enantiomer. Alternatively, an extraction procedure has been employed; this involves the interaction of an aqueous solution of the racemic guest with an immiscible organic solution of the optically-active host. The organic layer is then assayed for both the total amount of guest extracted and the proportion of each diastereoisomer present. Of course, the host must be sufficiently lipophilic to remain in the organic layer and the guest should not be able to be extracted in the absence of the host.

Host (232), incorporating *chiral* binaphthyl units, shows high chiral discrimination on interaction with particular racemic amine salts (Helgeson *et al.*, 1974). Such subtle behaviour mimics the discrimination that is characteristic of interactions between a wide range of optically active biological molecules. Chiral recognition of this type also provides a molecular basis for the design of new hosts for the practical resolution of amino acids and their esters (Peacock & Cram, 1976). Chiral crowns have been employed for optical resolutions via both liquid-liquid chromatography and solid-liquid chromatography (using the crowns immobilized on solid substrates such as silica gel or polystyrene).

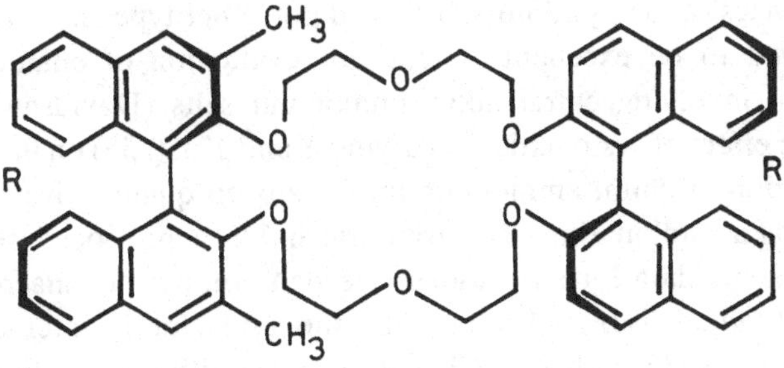

(232)

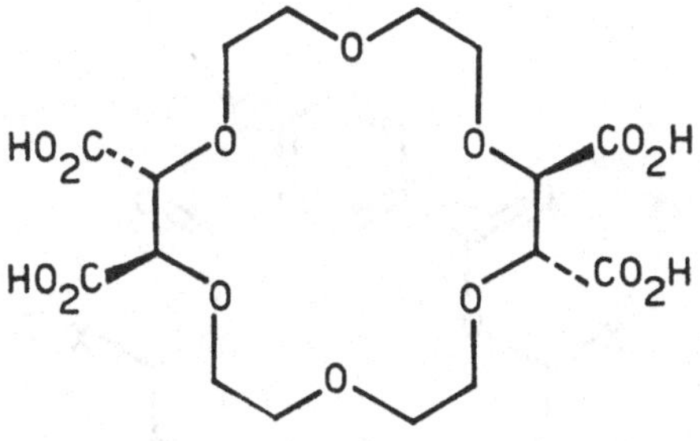

(233)

Chirality has also been introduced into crown hosts using optically-active functional groups other than bis-β-naphthol. For example, the crown (233) derived from L-tartaric acid is a chiral host showing much less

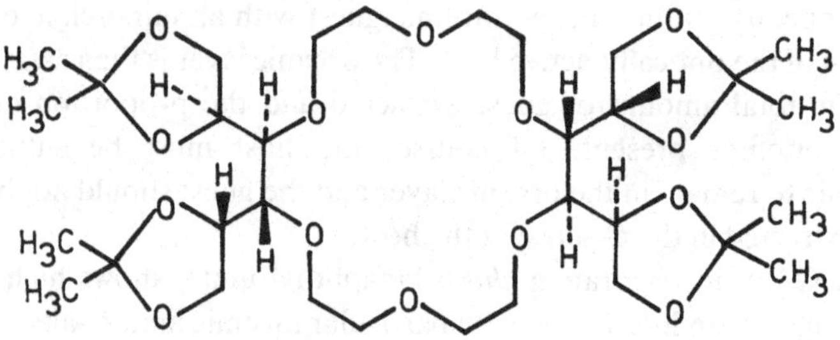

(234)

distortion of the periphery of the crown than occurs for (232) (Girodeau, Lehn & Sauvage, 1975; Behr, Lehn & Vierling, 1976). In other systems, optically-active sugar residues have been incorporated in the crown structure. The D-manitol derivative (234) is one example (Stoddart, 1979). These systems each exhibit varying degrees of chiral recognition towards a range of optically-active hosts.

Macrocycles of the pyridino-substituted polyether type, such as (235), have proved to be excellent systems for evaluation of enantiomeric recognition involving chiral alkyl ammonium salts (Davidson *et al.*, 1984). The effect of systematically varying X and Y in (235) on the extent of recognition of chiral ammonium ions has been quantitatively determined using a combination of calorimetric and ^1H nmr procedures. Full thermodynamic data for enantiomeric recognition by such macrocycles has been obtained. Table 5.1 summarizes the results for the interaction of the S,S-form of (235; with $X = CH_3$ and $Y = H$) with the enantiomers of two chiral alkyl ammonium salts. The data clearly confirm that significant chiral discrimination occurs for these systems.

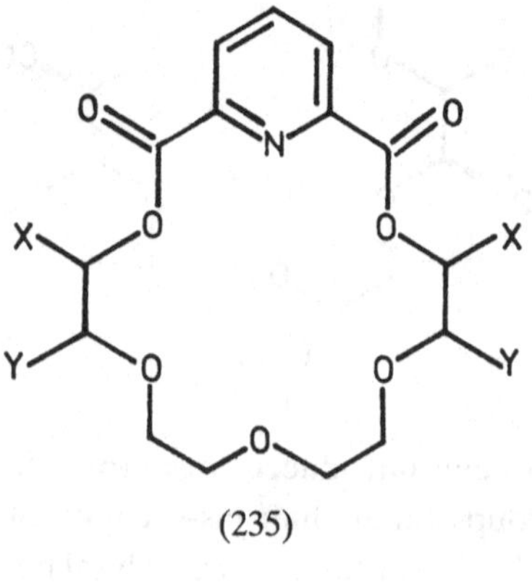

(235)

Table 5.1. Log K, ΔH (kJ mol^{-1}), and $T\Delta S$ (kJ mol^{-1}) values for the interaction of the S,S-isomer of (235; $X = CH_3$, $Y = H$) with the enantiomers of two chiral alkylammonium salts[a] (Davidson *et al.*, 1984).

	NapC$_2$H$_5$[b]		AlaOCH$_3$[c]	
	S	R	S	R
log K	2.06	2.47	1.78	2.02
ΔH	−26.44	−27.57	−14.56	−14.77
$T\Delta S$	−14.69	−13.47	−4.44	−3.26

[a] In CH$_3$OH at 25 °C.
[b] NapC$_2$H$_5$ = the hydrogen perchlorate salt of α-(1-naphthyl)ethylamine.
[c] AlaOCH$_3$ = the hydrogen chloride salt of methyl alaninate.

Complexes of neutral guests. As might be predicted, complex formation between uncharged crown ethers and neutral guests tends to be less strong than with cationic guests. Thus, although the complexes of a range of uncharged guests have been documented (Vogtle, Sieger & Muller, 1981), in many instances the guest is quite loosely bound and often lies outside of the cavity of the crown. In addition, there are many reports of complexes showing other than 1:1 stoichiometry. For example, part of the range of thiourea-crown complexes reported so far is given in Table 5.2; all these products have been obtained as crystalline solids (Pedersen, 1971).

Water has also been demonstrated to act as a guest in a number of crown complexes. Indeed, as a consequence of their ability to associate with water, the crowns have been demonstrated to be effective reagents

Table 5.2. A selection of host-guest complexes between crown polyethers and thiourea (Pedersen, 1971).

Crown host:	Guest	Stoichiometry
benzo-15-crown-5:	thiourea	1 : 4
benzo-18-crown-6:	thiourea	1 : 1
dibenzo-15-crown-5:	thiourea	1 : 1
dibenzo-18-crown-6:	thiourea	1 : 1
dibenzo-21-crown-7:	thiourea	2 : 7
dicyclohexyl-18-crown-6:	thiourea	1 : 6

for solubilizing water in apolar solvents such as chloroform (de Jong, Reinhoudt & Smit, 1976).

An interesting example in which the water is strongly bound (it is not lost from the complex at elevated temperatures) is given by (236). The X-ray structure of this complex indicates that the water molecule is included in the macrocyclic cavity with three strong hydrogen bonds binding it to the macrocyclic host as illustrated in (236) (Goldberg, 1978). The second –OH group in the host does not bind to the water but is involved in an intermolecular hydrogen bond.

(236)

Complex formation involving a range of other simple inorganic guests has been investigated. For example, in early work, the complex between Br_2 and dicyclohexyl-18-crown-6 was reported (Shchori & Jagur-Grodzinski, 1972). Subsequently, other crowns were demonstrated to form molecular complexes with bromine although the nature of many of these species remains to be elucidated.

The host-guest complexes of crowns with a range of polar organic guests have also been studied. Acetonitrile forms a crystalline complex with 18-crown-6 which has been used to purify this crown (Gokel *et al.*, 1974). Thus, colourless crystals may be isolated from a solution of 18-crown-6 in acetonitrile. Pure crown may then be recovered by removal of the bound acetonitrile using vacuum distillation. Related complexes of other volatile organic guests such as nitromethane, dimethylformamide, α-picoline, benzyl chloride and acetic anhydride have also been reported. A wide variety of other polar organic guests (including many substituted anilines, phenols, and hydrazines) may be used for formation of molecular complexes with crowns. A number of these complexes have application in phase transfer catalysis (reflecting the fact that complexation often markedly increases a guest's solubility in non-polar solvents). For other complexes, enhanced stability of particular guests is induced. An example of this is the observed stabilization of hydrazine reagents, such as phenylhydrazine, on complexation by crowns.

In the complex between the bis-methylester $CH_3OOCC \equiv CCOOCH_3$ and 18-crown-6, van der Waals forces appear to dominate the host-guest binding (Timko *et al.*, 1977). The X-ray crystal structure of this adduct indicates that each methyl group of the guest is 'perching' on a separate crown (Goldberg, 1975b). In addition to strong van der Waals contacts, two hydrogens from each methyl appear to be weakly hydrogen bonded to two crown oxygens; in this arrangement all oxygens of the respective crowns are orientated inwards.

Complexes of transition metal-ammines. In a novel extension of the areas just discussed, transition metal ammines have been shown to form adducts with either 18-crown-6 or its dibenzo derivative in both aqueous and non-aqueous solution (Colquhoun, Lewis, Stoddart & Williams, 1983). A number of such products, involving several transition metal ammine complexes, have been crystallized. In these unusual species, a bound ammine in the metal complex bridges to the crown host via three hydrogen bonds. Both monomeric and polymeric adducts have been isolated. Use of dibenzo-18-crown-6 appears to favour formation of discrete 1:1 species – in these the ammine guest interacts with only one face of the crown. The structure of the adduct formed between dibenzo-18-crown-6 and square planar $[PtCl_2P(CH_3)_3NH_3]$, is illustrated by (237). The resemblance between the three-point binding in this (and related species) to the binding in molecular complexes involving primary

(237)

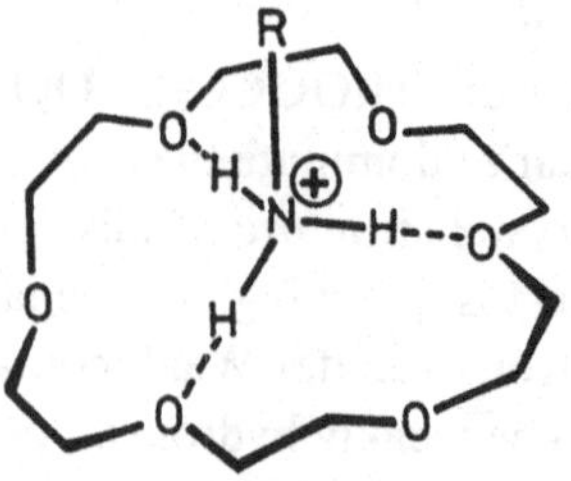

(238) 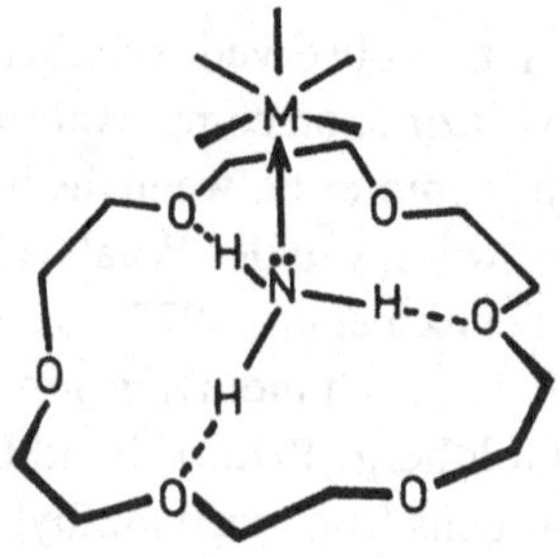   (239)

alkylammonium ion guests is evident on comparing (238) with (239) (illustrated here as the adduct of an octahedral metal).

Adduct formation of the present type has been shown to provide a potentially useful method for separation of transition metals. Thus, 18-crown-6 selectively precipitates the Cu(II) tetrammine complex (as a polymeric 1:1 adduct) in the presence of a corresponding concentration of Co(III) stabilized as its hexammine complex.

Cryptand and related cage hosts

As discussed in Chapter 4, the selectivity of cage ligands (such as the cryptands) for particular guests tends to be more readily controlled by structural modification than is the case for the crowns. This is usually a reflection of the cavities being inherently better defined in the three-dimensional cage structures.

Cationic, neutral and anionic guest complexation. Cage hosts incorporating non-metal cationic guests (such as the ammonium ion) have been documented. However, when protonated nitrogen atoms are present, cages such as the cryptands are often ideal for trapping anionic species in their cavities. The mode of host-guest binding in such complexes differs from that in the complexes of cationic and neutral guests discussed so far since the latter only involve electronegative donor sites in the host. In contrast, complexes incorporating anionic guests contain at least one positively charged receptor site. Hence the stability of such species depends not only on complementary host-guest steric properties but also on the 'reverse' charge neutralization which occurs on complex formation.

The macrotricyclic ligands (240) and (241) may be synthesized by multistep high-dilution procedures (Graf & Lehn, 1975). They contain 'spherical' cavities which are able to accommodate suitable guests whether they be cationic, neutral, or anionic.

(240)   (241)

Both these cages form complexes with most of the alkali and alkaline earth ions. In particular, the caesium complex of (240) is especially stable compared to other complexes of this ion.

Ligand (240) also forms a 1:1 inclusion complex with the ammonium ion. The latter is arranged in the cavity such that there appears to be a tetrahedral array of hydrogen bonds formed between the ammonium hydrogens and the nitrogen donor sites of the cage; the ether oxygens also appear to interact electrostatically with the central (charged) nitrogen atom. The inclusion of the ammonium ion by this host has been confirmed by X-ray diffraction (Metz, Rozalky & Weiss, 1976).

In its diprotonated form, (240) has been shown to include a water molecule. Once again, this guest is bound by an H-bonding array which involves two H-bonds from NH^+ sites on the host to the water oxygen; two further H-bonds form between the unprotonated amine sites and the water hydrogens – see (242). Using nmr, it was demonstrated that water exchange is slow for this species, reflecting the presence of tight host-guest binding.

(242)

In other studies, it has been shown that both tetraprotonated (240) and (241) form inclusion complexes with halide anions (Graf & Lehn, 1976). The chloride complexes of both these cages are quite stable with the log K values being approximately 10^4 in each case. These systems exhibit selectivity for chloride over bromide. The accommodation of a spherical halide anion by the tetrahedral bonding arrangement is illustrated by structure (243). The iodide ion (radius 3.57 Å) is too large for the cavities of (240) and (241) and does not form an inclusion complex with either of these cages. Similarly, the larger polyatomic anions NO_3^-, CF_3^-, ClO_4^- and $RCOO^-$ are not included.

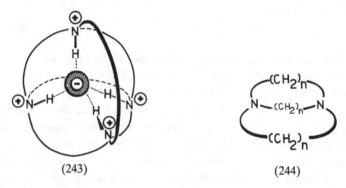

(243)   (244)

Diazabicycloalkanes of type (244), incorporating two bridgehead nitrogen groups, were synthesized in early studies (Park & Simmons, 1968). On protonation, these cages are also able to encapsulate halide ions, provided the cavity size is sufficient to accommodate the particular ion of interest. When all three chains connecting the nitrogen bridgeheads contain eight carbons, then chloride, bromide and iodide are all too large to be included. When the chains are nine atoms long, chloride or bromide may occupy the cavity but not iodide; with ten-atom chains, all three of these anions are now readily incorporated in the cavity. On complex formation, the cage assumes the *in–in* configuration shown in (245) in which the central halide ion is symmetrically bound by two H-bonds to the $-NH^+-$ groups of the diprotonated host.

(245)

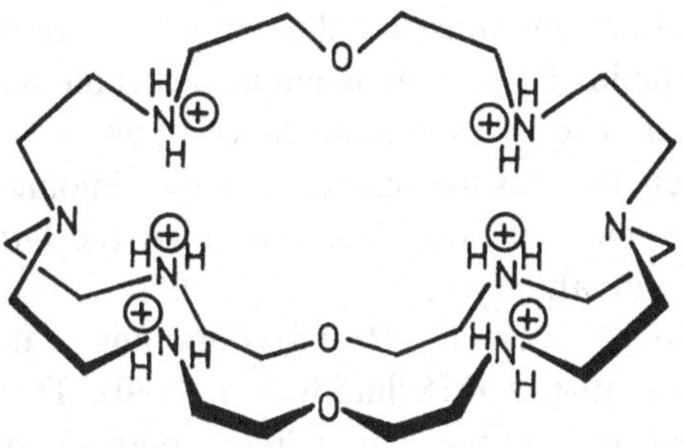

(246)

Larger cage hosts than those discussed so far have also been synthesized. Typical of these is the hexaprotonated cage derivative (246) of the 'bis-tren' type (Lehn, Pine, Watanabe & Willard, 1977). The cavity in this cage is able to accommodate spherical anions such as fluoride, chloride, and bromide and is also able to bind linear anionic species such as azide (Dietrich *et al.*, 1984) – see (247). It is interesting that, on inclusion, the small fluoride ion is four-coordinate while the larger chloride and bromide ions are six-coordinate. Ligand distortions occur in the structure of the latter complexes (Kintzinger *et al.*, 1983) and these appear to arise primarily from the inclusion of a spherical anion in an ellipsoidal cavity. The cylindrical triatomic azide ion (N_3^-) better matches the cavity shape. This guest is bound by two pyramidal arrays of three hydrogen bonds – each of which interacts with a terminal nitrogen from the azide ion. Complex formation by (246) and a variety of polyatomic anions has also been observed (Motekaitis, Martell, Lehn & Watanabe, 1982; Dietrich *et al.*, 1984). It is clear from these studies that this host is best considered to contain an ellipsoidal recognition site which thus contrasts with the spherical sites of the tricyclic derivatives (240) and (241) discussed previously.

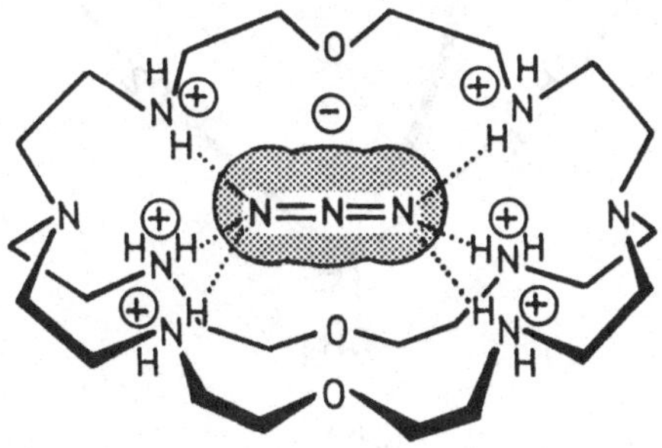

(247)

All these studies serve to illustrate the dominant importance of hydrogen bonding in stabilizing particular anion inclusion complexes of the type just discussed. The relative importance of any direct electrostatic attraction between the positive cavity and the anionic guest is more difficult to define. Nevertheless, interactions of the latter type are of crucial importance in other systems.

Symmetrical N_4-cages (248; $n = 6$ or 8) containing quaternary nitrogen atoms have been synthesized (Schmidtchen, 1980). These species bind halide ions strongly, even though it is no longer possible to form hydrogen bonds between the host and the halide. Clearly, such complexes are stabilized by direct electrostatic interaction between host and guest. Host (249) also yields related chloride and bromide complexes; however, unlike (248), (249) is still too small to accommodate the iodide ion.

The study of anion coordination has implications for a number of areas in chemistry and biochemistry. These include analytical applications concerned with anion sensing or separation as well as model studies for anion-specific biochemical systems. With respect to the latter, it is of interest that the majority of enzymic systems so far characterized bind substrates which are anionic.

(248)

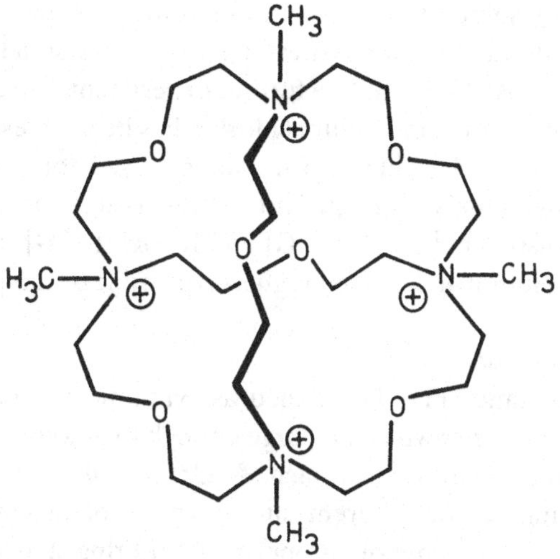

(249)

A system exhibiting chiral recognition. The chiral macrotricyclic tet-raamide (250) (Lehn, Simon & Moradpour, 1978) has been used for the complexation, extraction and transport of primary ammonium salts. The tetraamide was used rather than the corresponding tetraamine because of the lower basicity of the nitrogens in the former ligand. This avoids the possibility of proton transfer occurring from the primary ammonium substrates $R–NH_3^+$ used as guests. In a typical experiment, a solution of a primary ammonium salt, such as naphthylethyl ammonium or phenylalanine methylester hydrochloride in hydrochloric acid was

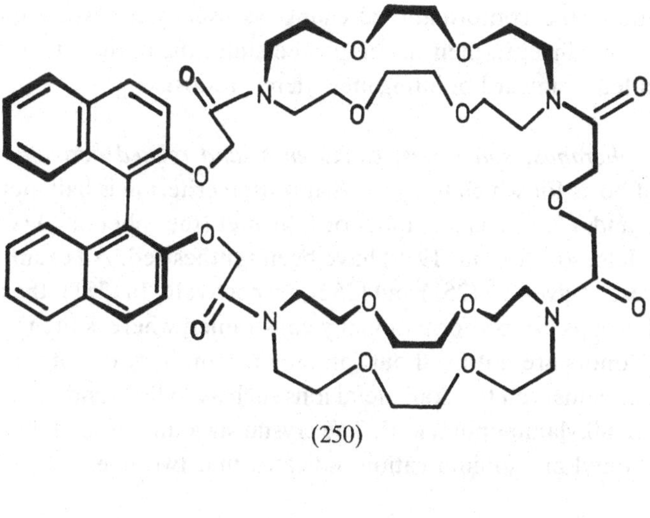

(250)

extracted with a solution of (250) in chloroform. Although the host contains two potential binding sites, an extraction stoichiometry no greater than 1:1 was observed. In a further experiment, the extraction of racemic α-naphthylethylammonium chloride has been investigated. The ^{1}H nmr spectrum of the chloroform solution used for the extraction shows two methyl signals of unequal intensities arising from the presence of the diastereoisomers [(−)**H**, (+)**G**] and [(−)**H**, (−)**G**], respectively. Thus partial resolution of the racemic guest is possible by this procedure.

Cavitand hosts

The cavitands may be defined as synthetic organic hosts that contain *enforced cavities* which are large enough to accommodate simple molecules or ions (Moran, Karbach & Cram, 1982). The cavitands encompass a number of different subcategories of macrocyclic hosts which thus have the common property of offering a fixed (or 'pre-organized') cavity to an incoming guest. The preparation of cavitand-type molecules has brought synthetic host-guest chemistry a step closer to the enzymic systems since, in the latter, the receptor sites also usually consist of fairly rigid cavities. Further, like the enzymes, the cavitands tend to involve a concave cavity which is organized to complement the normally convex surface of an included guest.

Early in the study of the host-guest chemistry of cavitands, it became apparent that increased complex stability arises because the cavities have been pre-organized for complexation during their synthesis rather than during their complexation step (Timko *et al.*, 1977). That is, rigid cavities in which the electron pairs point inwards are preferable to those for which a conformational change is necessary for this to occur. As already mentioned, such a conformational change is usually necessary for the simple crowns since, in their uncomplexed state, the cavities tend to be partially filled by inward-pointing methylene groups.

The hemispherands, spherands, calixarenes, and related derivatives. A number of hosts for which the pre-organization criterion is half met (the hemispherands) (Cram *et al.*, 1982) or fully met (the spherands) (Cram, Kaneda, Helgeson & Lein, 1979) have been synthesized. An example of each of these is given by (251) and (252), respectively. In (251), the three methoxyl groups are conformationally constrained whereas the remaining ether donors are not fixed but can either point in or out of the ring. This system binds well to alkali metal ions such as sodium and potassium as well as to alkylammonium ions. The crystal structure of the 1:1 adduct with the *t*-butyl ammonium cation indicates that two linear $^+$N–H$\cdots$O

(251)   (252)

bonds link this guest to two oxygens of the flexible ether bridge; the
remaining $^+$N–H hydrogen points towards the three aromatic –OCH$_3$
oxygens to yield what at first appears to be a trifurcated hydrogen bond
but is probably better described as a bifurcated H-bond (Cram & Trueblood, 1981).

The spherands [of which, collar-shaped (252) may be considered to be
the prototype] have been demonstrated to be highly selective and power-
ful complexing agents for particular alkali metal ions (Cram *et al.*, 1981;
Lein & Cram, 1982). For example, although (252) appears to contain
a cavity which is too small to readily complex most organic hosts, it
does bind strongly to lithium and sodium. However, the larger spherand
(253) is able to form stable complexes with a range of non-metal guests
such as $H_3NOH^+.H_2O$, $H_3NNH_3^{2+}$, $H_3N(CH_2)_2CH(CO_2H)NH_3^{2+}$ and

(253)

$1,3-H_3NC_6H_4NH_3^{2+}$; it also binds strongly to caesium (Helgeson, Mazaleyrat & Cram, 1981). For this spherand, the eight oxygens point alternatively above and below the mean donor atom plane such that they occupy the apices of a square antiprism. As might be predicted from its larger cavity size, the analogous spherand containing ten methoxy functions no longer complexes caesium strongly. Models indicate that cyclohexane in the chair conformation can be inserted into the cavity of this larger spherand and, on co-crystallizing it with cyclohexane, the solid 1:1 inclusion complex is obtained.

A number of other rigid host molecules have also been developed and, as before, strong complexation is observed when there is a good match between host and guest. Derivative (254), incorporating cyclic urea moieties, forms complexes with the alkali metal ions as well as with a range of ammonium ions (Nolte & Cram, 1984). The relative complex stabilities fall in the order: $Li^+ > Na^+ > K^+ > Rb^+ < Cs^+ < NH_4^+ < CH_3NH_3^+ < t-C_4H_9NH_3^+$. It should be mentioned that complex formation by a number of other urea-containing spherands has also been investigated in some detail (Katz & Cram, 1984).

(254)

A further category of cavitands are the calixarenes (Gutsche, Dhawan, No & Muthukrishnan, 1981; Gutsche & Levine, 1982). Structure (255) illustrates an example of this type which is readily prepared by treatment of 4-*t*-butylphenol with formaldehyde and base. The compound may exist in other conformations besides the saucer-shaped one illustrated by (255). Similarly, *t*-butyl-calix[4]arene (256; R = CH_2COOH) has an enforced hydrophilic cavity in the shape of a cone; the alkali and ammonium salts of this host are soluble in water (Arduini, Pochini, Reverberi & Ungaro, 1984).

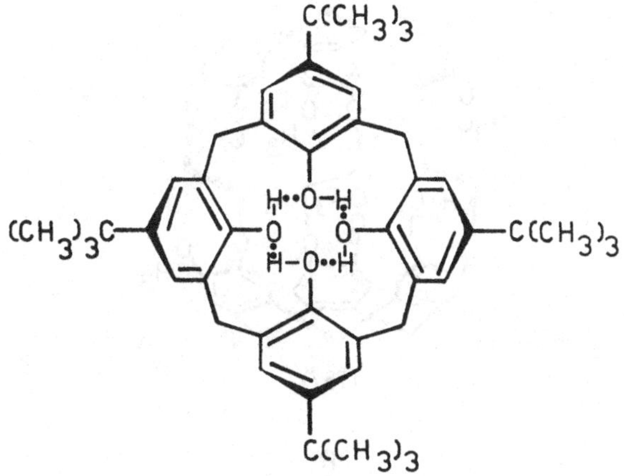

(255)

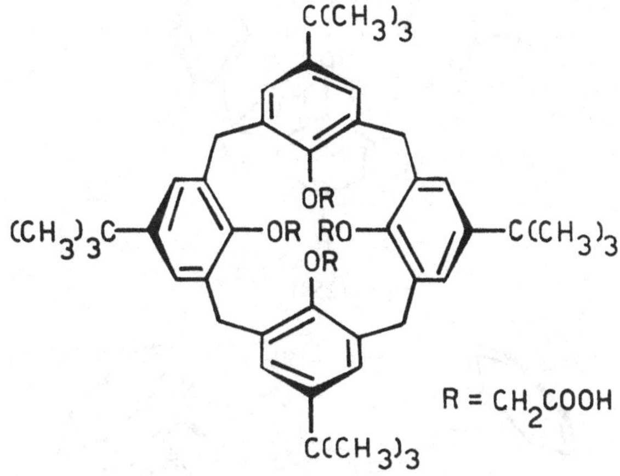

R = CH$_2$COOH

(256)

Further examples of cavitand-type structures include *bis*-cyclo-triveratrylene derivatives such as (257) (Gabard & Collet, 1981; Canceill, Lacombe & Collet, 1986) and the bowl-shaped hosts represented by (258) – the base of the bowl is formed by the four methyl groups. Once again, the shape of these molecules is maintained by conformational constraints. Cavitand (258) is able to accommodate simple solvent molecules such as dichloromethane and chloroform. Moreover, its cavity is large enough to form inclusion complexes with up to four molecules of water (Moran, Karbach & Cram, 1982).

(257)

(258)

(259)

Molecular models indicate that cavitand (259) has a tall, vase-shaped architecture in which the four benzene rings and four nine-membered rings combine to form a concave cavity to which is attached the four diazanaphthalenes (Moran, Karbach & Cram, 1982). The latter groups resemble flaps which may occupy either equatorial or axial positions. In the former arrangement, the inner surface area of the cavity is quite reduced and hence a considerable decrease in the inclusion ability of this form is expected.

The cryptahemispherands (Cram & Ho, 1986) form a 'developed' group of cavitands incorporating structural elements from each of the spherands, hemispherands, cryptands and simple crowns. Compound (260) typifies this group. As before, these hosts form complexes with a range of alkali metal ions as well as with the ammonium cation – often binding these guests more strongly than corresponding members from each of the related component categories.

(260)

Cram and co-workers have been successful in modifying certain of their cavitands such that reactions with a bound substrate are promoted. Such systems provide a first step towards the synthesis of rudimentary 'enzymes' (Cram, Katz, & Dicker, 1984). One example of this type, involving a binding step followed by a fast acylation step, is illustrated by Figure 5.1. This sequence resembles part of the mechanism used by chymotrypsin to cleave a peptide bond. Thus, the enzymic process entails several stages but, like the model system, begins with a binding step followed by a crucial transacylation step.

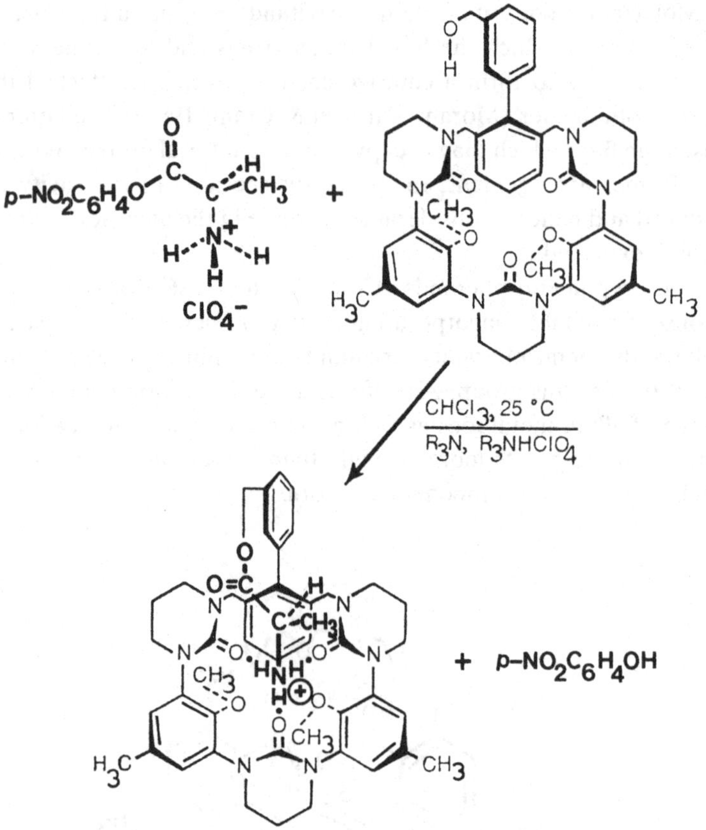

Figure 5.1. Fast acylation of alanine ester via host-guest complexation (from *Chemistry and Engineering News*, 1983, p. 33).

5.3   Host-guest complexation involving the cyclophanes

The cyclophanes are large-ring compounds incorporating benzene nuclei in the rings. In one sense, the cyclophanes may be considered to be a further category of the cavitands since the aromatic rings impart rigidity to the overall cyclic structure. The inclusion behaviour of cyclophanes has been the subject of a great deal of study.

In 1955, cyclophane derivatives of type (261); (with n = 3 or 4) were described (Stetter & Roos, 1955). After recrystallization from benzene or dioxan, these species crystallized as the 1:1 adducts of these solvents. Once such adducts are formed, it is difficult to remove the solvent and it was concluded that the solvent molecules are housed inside the cavities of the respective guests. In accord with this, the smaller ring system (261;

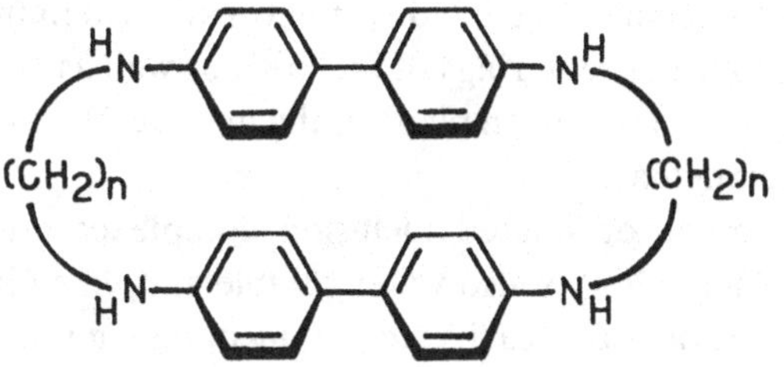

(261)

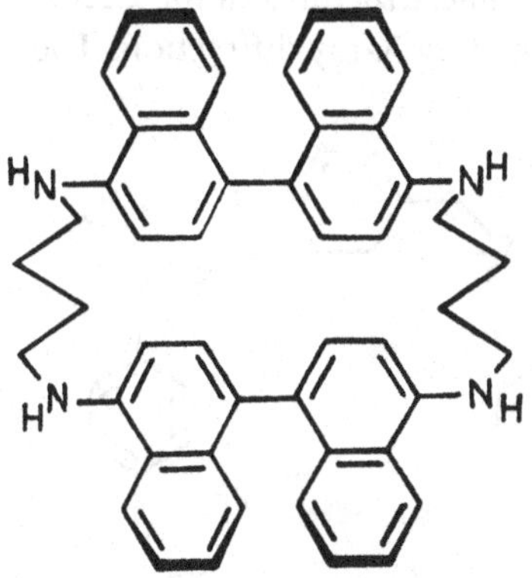

(262)

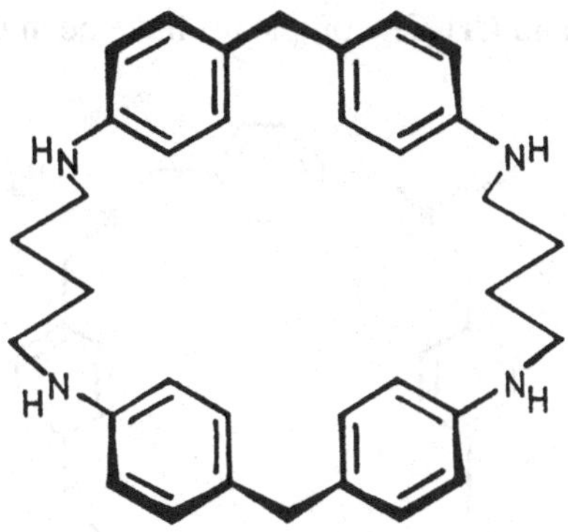

(263)

$n = 2$) is unable to form similar complexes with these solvents. In subsequent work, the heterocyclophane derivative (262), incorporating binaphthyl moieties, was synthesized. As expected, this guest also readily forms solvent adducts (Faust & Pallas, 1960).

More recently, the water-soluble paracyclophane (263) was demonstrated to form crystalline complexes with a range of hydrophobic substrates under acid conditions (Odashima, Itai, Iitaka & Koga, 1980). For example, with durene (264), a complex of stoichiometry [host.4HCl. durene.4H$_2$O] was obtained. The X-ray structure of this species indicates

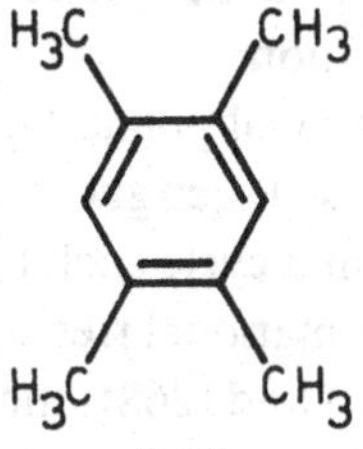

(264)

that a durene molecule is included symmetrically within the host's cavity such that the four benzene rings of the host, as well as that of the guest, are perpendicular to the mean plane of the macrocyclic ring; the cavity is effectively rectangular.

A large number of related inclusion complexes are now known. Heterocyclophane (265) is also water soluble and, like (263), presents a rectangular hydrophobic cavity to an incoming guest. The distance between each pair of parallel benzene side walls is about 5 Å (Tabushi, Kuroda & Kimura, 1976). This derivative forms 1:1 molecular complexes with $CHCl_3$, CH_2Cl_2, CH_2BrCl or CH_3CN. The structures of the $CHCl_3$ and CH_2Cl_2 complexes have been determined by X-ray diffraction. The

(265)   (266)

X-ray data confirm that (265) resembles a 'square box' in which the walls are formed by the benzene rings and the floor is the plane containing the four nitrogen atoms. The depth of the cavity so produced is about 6 Å and $CHCl_3$ fits in this cavity such that there is near-optimal van der Waals stabilization. The four-sulfur analogue (266) has also been prepared by condensation, in a benzene/ethanol mixture, of the appropriate p-substituted bis(methylthiolo) and bis(methylbromo) fragments, (Tabushi, Sasaki & Kuroda, 1976). In the form of its tetramethylsulfonium salt, (266), this cyclophane undergoes tight binding with a number of hydrophobic guests in aqueous solution.

Other rectangular or hexagonal phane-type cavitands have been synthesized. For example, a series of such guests consist of rigid dibenzofuran units bound to one another in a cycle such that the resulting macrocycle exhibits a low order of conformational freedom (Cram, 1983). Examples of this type are given by (267) and (268); the dimensions of the cavity in the larger ring species (268) are approximately 11 × 7 × 7 Å. Never-

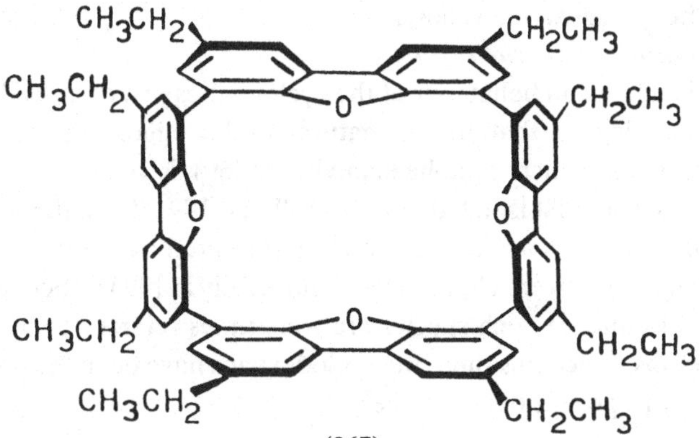

(267)

theless, it should be noted that the cavity of this larger-ring species is able to undergo partial 'collapse' such that the dimensions are reduced to approximately $11 \times 9 \times 3.5$ Å. As might be expected, both (267) and (268) yield crystalline complexes with a variety of solvents.

It should be noted that not all host-guest phane complexes are of the type just described. Indeed, a significant number of 'lattice inclusion' complexes also occur. In these, the host molecules stack such that a channel running between hosts is formed. Guest molecules occupy this channel. As expected, such an arrangement is usually reflected by relatively poor host-guest selectivity.

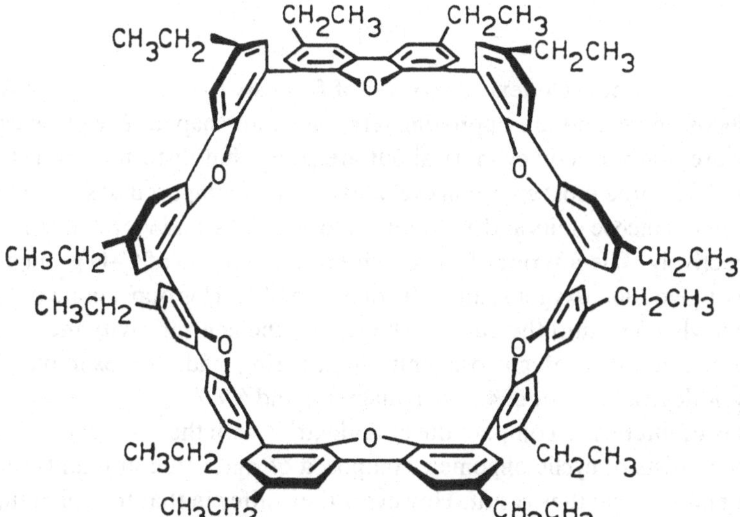

(268)

5.4   The cyclodextrins as hosts
Nature of the cyclodextrins

The inclusion behaviour of the cyclophanes just discussed parallels to some degree that of the natural cyclic oligosaccharides, the
cyclodextrins, of which cyclohexaamylase (269) is an example. The
cyclodextrins were first isolated in 1891 (Villiers, 1891) from the action of
amylase of *Bacillus macerans* on starch and related compounds. These
non-reducing sugars were characterized more fully in 1904 by Schardinger
and as a consequence such species are sometimes referred to as Schardinger dextrins. Since this time, the cyclodextrins have been the object of
an extremely large number of studies.

(269)

The family of cyclodextrins consist of from six to twelve α-1,4 linked
D-glucose units and are approximately doughnut shaped. Nevertheless,
they are not perfectly cylindrical but are somewhat distorted towards a
cone. The three most common cyclodextrins are composed of six, seven,
and eight glucose units and are referred to as, α-, β-, and γ-cyclodextrins,
respectively. The interior of each cavity contains a ring of C–H groups, a
ring of glycoside oxygens, and a further ring of C–H groups; none of the
hydroxyls point into the cavity. The size of the central cavity increases
with the number of glucose units in the ring and, for example, in
α-cyclodextrin it is about 4.5 Å in diameter and 6.7 Å deep.

It is of interest to compare the cyclodextrins with the spherands, since
each consists of cyclic oligomers composed of rigid structural units such
that enforced cavities result. However, they differ in that the spherands

tend to be highly lipophilic on the outside and somewhat hydrophilic on the inside, compared to the cyclodextrins which are very hydrophilic on their 'rims' but largely lipophilic on the remainder of their inner and outer surfaces. Unlike most spherands, the cyclodextrins are water soluble – a direct reflection of their hydrophilic outer edges.

Inclusion behaviour

The cyclodextrins have been demonstrated over many years to form inclusion complexes with a wide range of inorganic and organic guests (Bender & Komiyama, 1978; Szejtli, 1982). These include substituted aromatic compounds (α-cyclodextrin is just large enough to accommodate a benzene ring), hydrocarbons, iodine, bromine and a range of simple gases (under pressure) including chlorine, krypton and small alkanes. In many cases, these complexes exhibit a 1:1 stoichiometry with association constants (K) which are moderately high (10^2–10^4 dm^3 mol^{-1}).

When p-nitrophenolate is incrementally added to α-cyclodextrin in water, the ultraviolet spectrum of this anion changes such that successive spectra give rise to two isobestic points (Figure 5.2). Such behaviour is in accord with the formation of a 1:1 species. The spectral changes may be used for the direct calculation of K, which in this case was found to be approximately 10^4 dm^3 mol^{-1} (Cramer, Saenger & Spatz, 1967).

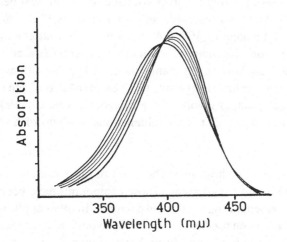

Figure 5.2. Spectral changes on incremental addition of p-nitrophenolate to α-cyclodextrin illustrating complex formation (from Cramer, Saenger & Spatz, 1967).

It should be noted that molecular complexes of the cyclodextrins may be isolated as crystalline solids; for example, a crystalline complex is obtained with iodine (which resembles the well known blue complex between iodine and starch) as well as with a large number of other inorganic and organic guests.

Once again, for strong complexation the guest molecule should fit the cavity well. Such a 'best fit' criterion has in fact formed the basis for the separation of cyclodextrins (Griffiths & Bender, 1973). Namely, cyclo-octaamylase can be coprecipitated with anthracene from aqueous solution while cycloheptaamylase and cyclohexaamylase remain in solution – anthracene is too large to form the necessary insoluble inclusion complexes in the latter cases.

The dynamics of complex formation and dissociation have been studied using a number of techniques. These include ^2H and ^{13}C nmr relaxation experiments (Behr & Lehn, 1976). The latter indicated that the inclusion complexes formed by α-cyclodextrin p-methylcinnamate, m-methylcinnamate, and p-$tert$-butylphenolate anions are weakly bound from a dynamic point of view. Upon complex formation, the overall tumbling motion of these three substrates is only slowed down by a factor of about 4. Thus, despite inclusion, these guests are quite dynamically labile – probably reflecting the lack of strong directional interactions in the binding between host and guest.

As before, the 'hydrophobic' bonding between cyclodextrins and organic guest molecules probably involves contributions from van der Waals interactions as well as favourable solvation changes (Tabushi, Kiyosuke, Sugimoto & Yamamura, 1978). A component of the latter may involve the expulsion from the cavity of 'enthalpy-rich' water (water in the cavity which does not have a full complement of hydrogen bonds compared to bulk solvent water) (Bergeron, 1977; Bender & Komiyama, 1978). Conformational changes involving the cyclodextrin on complex formation have also been postulated to influence the strength of host binding.

Cyclodextrin applications. The host-guest chemistry of the cyclodextrins has been exploited in a number of areas. These compounds have been commonly employed as separating agents and catalysts. In other applications, cyclodextrins have been used as stabilizers in flavour and fragrance chemistry and have been incorporated in oral medicines to control the rate of drug release or as masking agents for unpleasant tastes. Underlying such uses, is the ability of a cyclodextrin to modify a guest's properties and especially its solubility, volatility or chemical reactivity.

Cyclodextrins as catalysts and enzyme models

It has long been known that cyclodextrins may act as elementary models for the catalytic behaviour of enzymes (Breslow, 1971). These hosts, with the assistance of their hydroxyl functions, may exhibit guest specificity, competitive inhibition, and Michaelis–Menten-type kinetics. All these are characteristics of enzyme-catalyzed reactions.

There have been many studies of the reactions of simple cyclodextrin complexes. A frequent feature of such systems is that functional groups on the guests are held adjacent to hydroxyl groups on the cyclodextrin. Such proximity is expected to promote increased (intra-complex) reaction rates in much the same manner that hydroxyl groups at the active site of particular enzymes readily attack an adjacent (bound) substrate molecule. In the latter case, rate enhancements of the order of 10^6 or more are not uncommon. Thus, just incorporating a substrate in the cyclodextrin cavity may enhance its reactivity, although in practice such enhancements are usually very modest. For example, under conditions in which 72% of anisole is bound in the cavity of α-cyclodextrin, p-chlorination using hypochlorous acid proceeds 5.3 times faster than occurs in the absence of the cyclodextrin (Breslow & Campbell, 1969; 1971; Breslow, 1982). The reaction proceeds with enhanced regioselectivity since no chlorination in the orthoposition of anisole is possible when this guest occupies the cyclodextrin cavity. In the absence of cyclodextrin, anisole is chlorinated at both ortho and para positions (see Figure 5.3). Anisole may also be chlorinated biologically by the enzyme chlorinase; however this enzyme is not as selective as the model system and gives rise to a mixture of both chloro derivatives!

Similarly, cyclodextrin accelerates the cleavage of pyrophosphates by about 200-fold. This enhancement is associated with a simultaneous transfer of a phenylphosphate group to the host by the vicinal action of the hydroxy groups [see Figure 5.4] (Hennrich & Cramer, 1965). In this case the product monophenylphosphate also forms an inclusion complex and thus 'product inhibition' occurs. Because of this, the system is not truly catalytic.

An additional example is the observed moderate acceleration in the cleavage of particular phenyl esters in the presence of a cyclodextrin. In such cases, the bound ester is attacked by an hydroxyl group on the cyclodextrin to yield a new ester. There was found to be a significant enhancement of phenol release from meta-substituted phenyl acetate on interaction with cyclodextrin (relative to other esters which do not fit the cavity so well) (Van Etten, Clowes, Sebastian & Bender, 1967). During the reaction, the acyl moiety transfers to an hydroxyl group on the

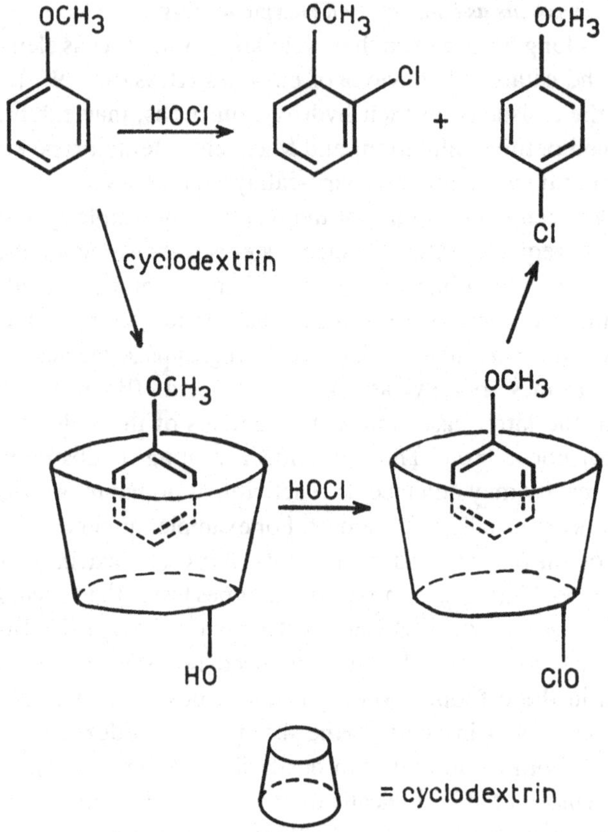

Figure 5.3. The selective chlorination of anisole in the presence of cyclodextrin (Breslow, 1982).

cyclodextrin. The mechanism appears to be formally similar to the chymotrypsin-catalyzed hydrolysis of esters. The rate of attack of the cyclodextrin hydroxyl group on the ester may be compared with the rate for solvent water under similar conditions in the absence of cyclodextrin. For a range of ester guests, the rate enhancement was greatest at 250-fold for *m*-tert-butylphenyl acetate but often less than 100-fold for other ester derivatives. A summary of the proposed mechanism is given by Figure 5.5 (Saenger, 1980).

Structurally developed cyclodextrins. Effective procedures for the selective functionalization of peripheral hydroxyl groups on the cyclodextrins have been developed. A motivation for these studies has been to produce suitably functionalized hosts which will induce enhanced reaction rates

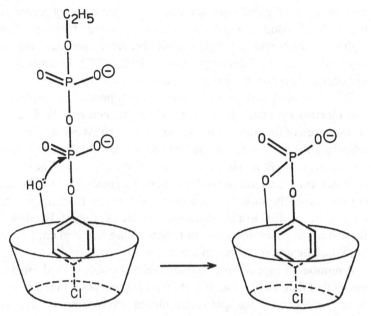

Figure 5.4. Schematic illustration of the cyclodextrin catalysed fission of a pyrophosphate derivative (Hennrich & Cramer, 1965).

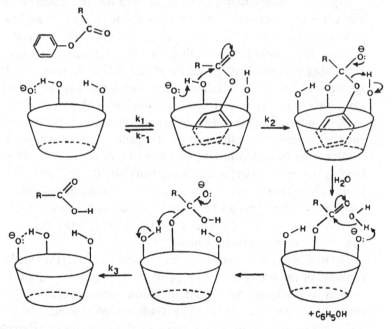

Figure 5.5. The cyclodextrin catalysis of phenylester hydrolysis (Saenger, 1980).

with an included guest while simultaneously maintaining geometric control of the reaction. Using this strategy it was hoped that the remarkably high selectivities and very high accelerations characteristic of the enzymes might ultimately be achieved (Breslow, 1982; 1983). In general, this has not been fully realized for the reactions discussed here.

Both flexible and rigid caps have been appended to one end of the cyclodextrin cylinder. These caps serve to increase the hydrophobic surface area of the host in contact with an included guest. This normally leads to an enhancement of the strength of binding of an included guest (sometimes by as much as an order of magnitude). These modified cyclodextrins also tend to hold the included guest more rigidly and, as a consequence, the modifications may result in the freezing of rotational degrees of freedom in a given transition state (Trainor & Breslow, 1981). In favourable cases, this may contribute to the required enhancement of both the rate and specificity of a particular reaction.

A number of capped cyclodextrins which are able to bind a metal ion at one end of their cavity, together with an organic guest in the cavity, have been synthesized. Such species parallel in several respects the family of completely synthetic 'vaulted' transition-metal complexes prepared by Busch and coworkers and already discussed in Chapter 3 (section 3.5).

In part, the attachment of a metal ion to a functionalized cyclodextrin was carried out in order to provide a model for substrate binding in metalloenzymes. Such derivatives were also synthesized in an attempt to increase the generally low catalytic activity associated with non-metal-containing cyclodextrins (Tabushi *et al.*, 1977; Tabushi, 1984). One system investigated is the adamantan-2-one-1-carboxylate adduct of the zinc-containing β-cyclodextrin derivative (270). Other derivatives similar to (270), but containing Cu(II), Zn(II), or Mg(II) have also been prepared and their interaction with a variety of organic guests investigated. These studies have provided insight into the role of a metal in aiding guest binding within the cavity of a cyclodextrin host. Triethylenetetraamine as well as diethylenetriamine have both been used to construct such flexible metal-containing caps; hosts of this type tend to hold their hydrophobic guests much more strongly than does the parent functionalized cyclodextrin containing no metal coordination.

The hydrolysis of esters by the nickel derivative (271) provided an early example of the use of a metal-capped cyclodextrin as a catalyst (shown here as its *p*-nitrophenyl acetate inclusion complex) (Breslow & Overman, 1970; Breslow, 1971). The synthesis of this host involves the following steps: (i) covalent binding of the pyridine dicarboxylic acid moiety to cyclodextrin, (ii) coordination of Ni(II) to this species, and (iii)

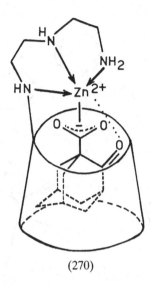

(270)

addition of 2-pyridine carboxaldoxime as a second ligand for the Ni(II). In the case where *p*-nitrophenyl acetate is the included guest, a rate enhancement for hydrolysis of greater than 10^3-fold was observed. The mechanism for this reaction appears to involve coordination of the guest in the cyclodextrin cavity such that the ester group is close to the oxime group of the metal-containing cap. An attack on the acetate group by the deprotonated oxime group occurs, resulting in loss of the *p*-nitrophenyl group. In a subsequent reaction, the methyl-acetylated oxime product is catalytically hydrolyzed with the aid of the attached Ni(II) ion.

(271)

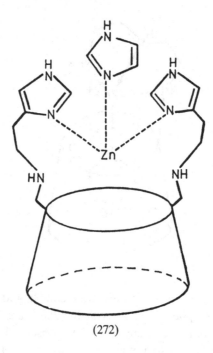

(272)

The system illustrated by (272) forms the basis of a model for the zinc-containing metalloenzyme, carbonic anhydrase (Tabushi & Kuroda, 1984). It contains Zn(II) bound to imidazole groups at the end of a hydrophobic pocket, as well as basic (amine) groups which are favourably placed to interact with a 'substrate' carbon dioxide molecule. These are both features for the natural enzyme whose function is to catalyze the reversible hydration of carbon dioxide. The synthetic system is able to mimic the action of the enzyme (although side reactions also occur). Nevertheless, the formation of bicarbonate is still many orders of magnitude slower than occurs for the enzyme.

A number of other 'structurally developed' cyclodextrins have been synthesized. For example, hosts containing two hydrophobic binding sites, called 'duplex' cyclodextrins, such as (273) (containing either α- or

(273)

β-cyclodextrins) have been prepared for use in catalytic studies (Tabushi, Kuroda & Shimokawa, 1979; Fujita, Ejima & Imoto, 1984).

Using a range of modified cyclodextrins related to the types just discussed, it has also proved possible to construct models for other enzymes; these include: ribonuclease, transaminase, lipase, and rhadopsin (Breslow, Hammond & Lauer, 1980; Tabushi, Kuroda & Mochizuki, 1980; Tabushi, Kimura & Yamamura, 1982). While many of these models are still fairly rudimentary, it is likely that synthetic systems, which approach the rates and selectivities of the enzymes, may well be increasingly produced in the future. Nevertheless, success in this area will depend to a large part on the ability to modify cyclodextrins so that favourable structuring of the guest in the binding site is engendered. It appears that achievement of this will largely overcome an inherent weakness in the majority of the models so far studied.

6

Thermodynamic considerations

6.1 Techniques for obtaining thermodynamic data

Thermodynamic aspects of the interaction of metal ions with macrocyclic ligands have been well studied. In many instances such studies have involved a comparison of the behaviour of cyclic ligand systems with that of their open-chain analogues. In this manner, information concerning the thermodynamic consequences arising from the *cyclic* nature of the macrocyclic ligand has been obtained. Frequently these studies have been restricted to stability constant (log K) measurements and, for such studies, a variety of techniques has been employed (Izatt *et al.*, 1985).

For ligand systems containing an ionizable proton (or containing sites which are capable of being protonated), classical potentiometric (pH) titrations have been widely used. Normally, the pK_a values for the ligand involved are first determined independently. Titrations are then performed on a solution containing the ligand and metal ion of interest. The procedure depends on competition of protons and metal ions for the ligand binding sites; the pH of the solution is monitored at each titration point after equilibrium has been established. From the known ligand pK_a values and the effect of the metal ion on the ligand titration curve it is possible to calculate the required stability constants. For 1:1 complexation, K is given by (charges not shown):

$$M + L \overset{K}{\rightleftharpoons} ML \qquad\qquad [6.1]$$

$$\text{where } K = \frac{[ML]}{[M][L]}$$

When the value for K is small, or the ligand system is otherwise not suitable for pH titration, then ion-selective electrodes have frequently been employed to determine the concentration of free metal ion in the metal-ligand solution after equilibrium has been established. Since the initial (that is, total) metal-ion and ligand concentrations are known, it is then usually straightforward to calculate K.

A wide variety of other methods have been used to obtain the concentration of either the metal or ligand in a particular equilibrium solution. These include polarography, spectrophotometry, magnetic resonance, cyclic voltametry and conductivity. Of course, if reliable results are to be obtained, the technique chosen must not significantly perturb the position of equilibrium in the solution.

As has been mentioned previously, the approach to equilibrium can often be slow for macrocyclic complex formation; indeed, equilibrium may take days, weeks or even months to be established. This may give rise to experimental difficulties in conventional titration procedures. Under such circumstances, it is usually necessary to carry out 'batch' determinations in which a number of solutions, corresponding to successive titrations points, are prepared and equilibrated in sealed flasks. The approach to equilibrium of each solution can then be monitored at will.

If the stability constant measurements are coupled with enthalpy determinations, then entropy values (as well as the corresponding free energy values) may be calculated:

$$\Delta G = -RT\ln K = \Delta H - T\Delta S. \qquad [6.2]$$

Two general procedures have been used to obtain ΔH values. The first involves the measurement of log K values over a range of temperatures; the observed variation may be used to derive the required ΔH value. However, because of the usual errors inherent in log K determinations coupled with the limited temperature range normally possible, ΔH values obtained in this manner tend to be somewhat unreliable. In contrast, the direct determination of ΔH using calorimetry commonly results in values which are considerably more accurate. Nevertheless, such calorimetric determinations may still not be easy for particular macrocyclic systems. Difficulties can arise in measuring the total heat evolved for metal complexation when long equilibration times are necessary. To lessen such problems, sensitive calorimeters have been used which are able to integrate the heat released over an extended time.

In some instances, a calorimetric titration procedure may be used to obtain both log K and ΔH values simultaneously. Thus, the titration curve (heat change versus volume of reactant added) gives a measure of the degree of complex formation at each titration point and, provided complex stability is not too high, may be used to determine the required K value directly. The total observed heat change is proportional to the ΔH value for the overall reaction being studied.

6.2   Macrocyclic effect

In pioneering work, it was demonstrated that the stability constant for the Cu(II) complex (red isomer) of the reduced Curtis macrocycle (274; tet-a) is approximately 10^4 times higher than for the related complex of the open-chain tetramine, '2.3.2-tet' (275) (Cabbiness & Margerum, 1969; 1970). Although it is expected that the stability of a particular complex type increases as the number of chelate rings increases (the **chelate effect**), the additional stability of the macrocyclic copper complex was about an order of magnitude greater than expected solely from the presence of an additional chelate ring. In view of this, Cabbiness and Margerum coined the term **macrocyclic effect** to describe this additional (unexpected) stability.

(274)   (275)

Origins of the effect

The thermodynamic origins of the macrocyclic effect, especially involving complexes of tetraaza ligands, have been the subject of many investigations. Early studies were contradictory and, for example, for particular systems the additional stability was assigned either wholly to entropy factors (Kodama & Kimura, 1976) or wholly to enthalpy factors (Hinz & Margerum, 1974). However, these opposite conclusions were based on studies in which the temperature dependence of the respective stability constants had been used to derive ΔH. This uncertainty concerning the relative importance of entropy versus enthalpy was largely resolved (for N_4-donor systems) by subsequent calorimetric determinations of the ΔH values for a range of open-chain and macrocyclic systems of this type. Typically, the studies were performed in aqueous media under strongly basic conditions where the metal ion (usually nickel or copper) is present as the hydroxo species and the strongly basic tetraamine ligands are not protonated (Bianchi *et al.*, 1984). This avoids the complication of protonation equilibria being present during the measurements.

Experimentally, the enthalpy change associated with the formation reaction

$$M^{2+}(aq) + L(aq) \rightleftharpoons ML^{2+}(aq) \qquad [6.3]$$

is determined calorimetrically for both the macrocyclic and open-chain systems. The difference between the respective enthalpy values gives the macrocyclic enthalpy for the metathetical reaction

$$ML^2(aq)^{2+} + L^1(aq) \rightleftharpoons ML^1(aq)^{2+} + L^2(aq). \qquad [6.4]$$

where L^1 is the macrocycle and L^2 is the open-chain analogue. From a range of such studies involving nitrogen donor ligands, it is clear that the entropy term associated with the macrocyclic effect tends to be favourable while the enthalpy term can be quite variable (and either favourable or unfavourable).

Table 6.1 summarizes the thermodynamic parameters relating to the macrocyclic effect for the high-spin Ni(II) complexes of four tetraaza-macrocyclic ligands and their open-chain analogues (the open-chain derivative which yields the most stable nickel complex was used in each case) (Micheloni, Paoletti & Sabatini, 1983). Clearly, the enthalpy and entropy terms make substantially different contributions to complex stability along the series. Thus, the small macrocyclic effect which occurs for the first complex results from a favourable entropy term which overrides an unfavourable enthalpy term. Similar trends are apparent for the next two systems but, for these, entropy terms are larger and a more pronounced macrocyclic effect is evident. For the fourth (cyclam) system, the considerable macrocyclic effect is a reflection of both a favourable entropy term and a favourable enthalpy term.

Before considering further the thermodynamic nature of particular macrocyclic effects, it is necessary to consider the various components of a typical complexation reaction. These are best illustrated by the Born–Haber cycle illustrated in Figure 6.1. Each of the steps 1–5 has a ΔG, ΔH and a ΔS term associated with it and the overall values of these parameters reflect the respective sums of the individual components. To understand fully the nature of a particular macrocyclic effect, it is necessary to have data available for each of these steps for both the macrocyclic system and open-chain (reference) system. Although some progress has been made

Table 6.1. *Parameters illustrating the macrocyclic effect for the high-spin Ni(II) complex of the tetraaza macrocycles L^2, L^4, L^6 and L^8 (Micheloni, Paoletti & Sabatini, 1983).*

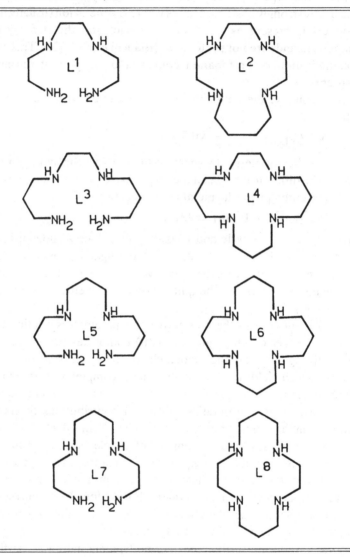

$$[NiL_{oc}]^{2+} + L_{mac}^{2+} \rightleftharpoons [NiL_{mac}]^{2+} + L_{oc}$$

where oc = open chain; mac = macrocylic.

L_{oc}/L_{mac}	$\dfrac{-\Delta G}{\text{kJ mol}^{-1}}$	$\dfrac{\Delta H}{\text{kJ mol}^{-1}}$	$\dfrac{T\Delta S}{\text{kJ mol}^{-1}}$
L^1/L^2	2.43	5.1	7.4
L^3/L^4	21.05	5.3	26.4
L^5/L^6	15.69	3.5	19.2
L^7/L^8	33.67	-20.5	13.2

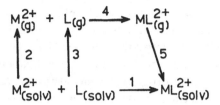

Figure 6.1. Born–Haber cycle for complex formation.

towards determination of a comprehensive set of thermodynamic data for particular systems (Clay, Micheloni, Paoletti & Steele, 1979), the general absence of unambiguous data for each of the steps has been the chief difficulty in defining precisely (as opposed to speculatively) the microscopic nature of particular macrocyclic effects.

The thermochemical cycle given in Figure 6.1 serves particularly to illustrate the critical role that solvation may play in particular cases. For example, it has been argued that desolvation of the metal ion and ligand (steps 2 and 3) will result in positive ΔH and ΔS changes. Thus, if it is assumed that similar desolvation of the metal ion occurs on its reaction with either the macrocyclic or open-chain ligand, then it is the desolvation of the respective ligands which is expected to differ and hence contribute to the macrocyclic effect. It has been further argued that less solvation will occur for the cyclic ligands than for their open-chain analogues because the former are more compact. If this is the case, then ligand desolvation enthalpies would be expected to be less for complexation of the macrocyclic ligand contributing to enhanced thermodynamic stability for the cyclic system (relative to the open-chain system).

The variability of the enthalpy term appears to reflect a number of influences. An important concern is any difference in the nature (and energy) of bonds between the metal ion and the respective ligands involved. Changes in ligand conformation on complexation will also be important as will the match or otherwise of the macrocyclic cavity for the metal ion. As just discussed, the different solvation energies of the open-chain and macrocyclic ligands may also play a crucial role in determining the enthalpic contribution (Clay, Micheloni, Paoletti & Steele, 1979). Thus it is stressed that there is a range of (often interdependent) factors which may influence the magnitude of this term for a particular complexation reaction and caution needs to be exercised when attributing an observed result to a single cause.

Considering their generally similar dimensions, there are not expected to be major differences between the solvation of the open-chain and macrocyclic metal complexes arising from size differences. Nevertheless.

a variation in the solvation term will occur if solvent interacts more strongly with the two terminal $-NH_2$ groups usually present in the open-chain reference complex than with the corresponding secondary amines in the macrocyclic complex. In order to appraise the suitability or otherwise of simple linear amines (incorporating terminal primary amine functions) as reference systems, a comparative study involving the corresponding ligands incorporating terminal di-*N*-methylated amine groups was undertaken. For two studies, the calorimetry results obtained were similar when the terminal $-NH_2$ groups in the reference ligand were changed to $-NHCH_3$ groups; this suggested that an effect of the latter type is not a major contributor to the macrocyclic enthalpy (Clay, McCormac, Micheloni & Paoletti, 1982). Nevertheless, for other systems, it is clear that the *N*-methylated derivatives yield solvation patterns which more closely resemble those of the macrocyclic systems (Clay, Corr, Micheloni & Paoletti, 1985). Thus, the behaviour of these *N*-methylated ligands towards both protonation and metal-complexation parallels more closely the behaviour of the tetraaza macrocyclic analogues. An interesting result from this study is that, when measured against the appropriate *N*-methylated open-chain reference system, the macrocyclic effect for the Ni(II) complexes of simple N_4-donor macrocycles involves both free energy and enthalpy terms which are essentially independent of ring size.

Because it also reflects the sum of a number of separate components, the entropic contribution to macrocyclic complex stability is similarly difficult to analyse in terms of specific effects. Thus changes in: (i) the total number of species present, and in (ii) the translational entropy of the system must be considered to be potentially important contributors to the total entropy. In particular, as already implied, the entropic consequences of solvation differences between the respective open-chain and cyclic species present are undoubtedly of major importance for particular systems. Further, the cyclic ligand will tend to undergo much less geometrical change on coordination than will its open-chain analogue. Hence, there will tend to be less 'loss of disorder' in the macrocyclic case with the occurrence of a more favourable 'configurational entropy' contribution to the overall thermodynamic stability. That is, the open-chain reference ligand is expected to have a considerable 'configurational' component contributing to its internal entropy because of its inherent flexibility; on coordination, a significant proportion of this component will be lost. In contrast, the less flexible macrocycle will not undergo such a marked change on coordination and its internal entropy will not be affected to the same degree.

Detailed analysis of a particular macrocyclic effect in the manner discussed so far is usually only justified if the open-chain and macrocyclic ligand species adopt similar coordination geometries: caution needs to be exercised since this is not necessarily the case. This is especially so when flexible ligands are involved and when the metal ion also has a 'flexible' coordination shell such as occurs, for example, with most non-transition ions (Micheloni & Paoletti, 1980).

Sulfur-containing ligands. Macrocyclic effects have also been documented in mixed donor systems. The Cu(II) complex of the 14-membered (*cis*) N_2S_2-donor analogue of cyclam exhibits a substantial macrocyclic effect with a log K difference of 4.6 relative to the corresponding open-chain species (Figure 6.2) (Micheloni, Paoletti, Siegfried-Hertli & Kaden, 1985). The effect in this case is mainly due to a favourable

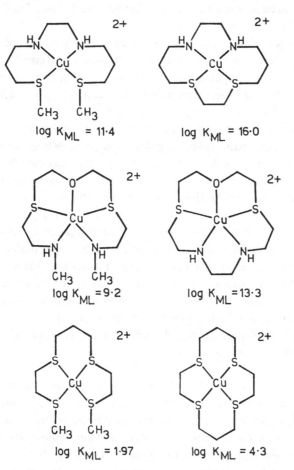

Figure 6.2. A comparison of the stabilities of the Cu(II) complexes of open-chain and macrocyclic ligands: values are for water at 25 °C with $I = 0.1$ or 0.5 mol dm^{-3}.

entropic term ($T\Delta S = 17.6$ kJ mol^{-1}) although there is also a lesser contribution from the enthalpic term ($-\Delta H = 8.4$ kJ mol^{-1}).

The Cu(II) complex of the 15-membered ON$_2$S$_2$-donor macrocycle shown in Figure 6.2 also shows a substantial macrocyclic effect (Arnaud-Neu, Schwing-Weill, Louis & Weiss, 1979). The origins of the effect in this case have been shown to be equally enthalpic and entropic. For this ligand system the macrocyclic effect is strongly metal-ion dependent: large effects occur for Cu(II), Ni(II) and Co(II) while only moderate effects are apparent for Cd(II) and Ag(I). In the case of Pb(II) or Zn(II), no macrocyclic effect is obvious – perhaps reflecting substantially different coordination geometries in the macrocyclic and reference complexes for these metals. Related metal-ion dependence of the effect has been demonstrated to occur for a number of other systems (Thom, Shaikjee & Hancock, 1986).

Macrocycles containing only thioether donor atoms have proved useful for probing the nature of the macrocyclic effect, since metal complexation is not complicated by competing protonation equilibria, and the generally lower polarity of these ligands (relative to their aza analogues) will tend to result in ligand solvation being less important.

Thermodynamic data for the reaction in aqueous solution of Cu(II) with a series of open-chain and cyclic polythioethers have been obtained. In general, the complexes show a significantly smaller macrocyclic effect than do the related tetraamine species. The Cu(II) complex of the 14-membered S$_4$-macrocycle (Figure 6.2) is only slightly greater than a hundred times more stable in water than the corresponding open-chain complex (Diaddario *et al.*, 1979; Sokol, Ochrymowycz & Rorabacher, 1981); although this difference is enhanced somewhat in methanol/water solvent mixtures. As determined by the temperature-dependence of the respective stability constants, the ΔH values obtained for the open-chain and macrocyclic systems were found to be near identical. Thus, the observed macrocyclic effect appears to be attributable solely to more favourable entropy – perhaps largely reflecting a favourable 'configurational' entropy component.

Crown polyethers. Macrocyclic effects involving complexes of crown polyethers have been well-recognized. As for the all-sulfur donor systems, the study of the macrocyclic effect tends to be more straightforward for complexes of cyclic polyethers especially when simple alkali and alkaline earth cations are involved (Haymore, Lamb, Izatt & Christensen, 1982). The advantages include: (i) the cyclic polyethers are weak, uncharged bases and metal complexation is not pH dependent; (ii) these ligands readily form complexes with the alkali and alkaline earth cations which can be considered to be simple solvated spheres free of stereo-chemical preferences; (iii) complex formation is normally quite rapid so that equilibrium is attained quickly; and (iv) stability constants tend to fall in the log K range 2–8 (in methanol) making them readily measured by direct means.

A stability enhancement in methanol of about 10^4 occurs for the potassium complex of 18-crown-6 (169) compared to its non-macrocyclic analogue (223) in methanol (Frensdorff, 1971; Petersen & Frensdorff, 1972) (Figure 6.3). A comparative study of this and the corresponding

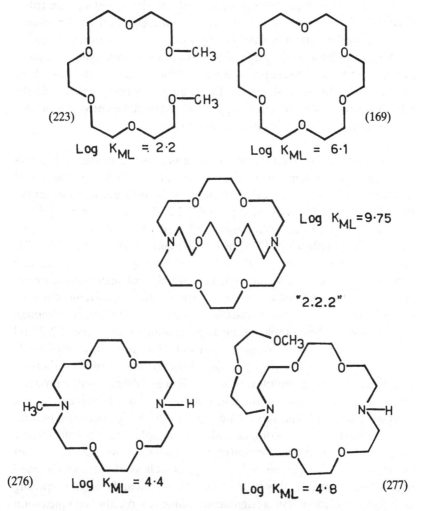

Figure 6.3. Stability constants for the 1:1 potassium complexes of some polyether ligands at 25 °C. The values for (223) and (169) were determined in methanol; those for 2.2.2, (276) and (277) in methanol/water (95:5).

sodium and barium systems was carried out (Haymore, Lamb, Izatt & Christensen, 1982). Using calorimetric titrimetry in both methanol and methanol/water mixtures, log K, ΔH and ΔS values for the respective systems were determined. The macrocyclic effects are manifested by a variation of between 2.9 and 4.5 orders of magnitude between the corresponding open-chain and macrocyclic stability constants. In contrast to the studies so far discussed in this chapter, the macrocyclic effect in these systems is consistently the result of favourable enthalpic factors – in general, entropic factors make very little contribution to the effect. Thus, for systems of this type, the expected 'configurational' entropy contributions are not evident; if configurational entropy changes are indeed contributing their effects are masked by other thermodynamic influences.

There have been a very large number of other thermodynamic studies involving the interaction of crowns with a wide range of metal-ions (Izatt *et al.*, 1985). In general, as for the examples already discussed, the complexes are found to be enthalpy-stabilized with the entropy term also contributing to the stability in a number of cases.

Bicyclic systems. An even greater enhancement of thermodynamic stability is evident on comparing the values for the metal complexes of macrobicyclic polyethers with their monocyclic analogues. For example, the potassium complex of the cryptand 2.2.2 (213; $m = 1$, $n = 1$) shows an approximate 10^5-fold increase in stability relative to the corresponding monocyclic complexes of (276) or (277) (Figure 6.3) (Frensdorff, 1971; Dietrich, Lehn & Sauvage, 1973; Lehn & Sauvage, 1975). As already mentioned in Chapter 4, the enhanced stability of such bicyclic ligand systems has been named the 'cryptate effect'. Once again it has not been possible to elucidate the microscopic origins of this effect, although comparison of the complexation of potassium by cryptand 2.2.2 and dicyclohexyl-18-crown-6 suggests that enthalpic terms are responsible (Kauffmann, Lehn & Sauvage, 1976). Further, comparison of thermodynamic data for the complexation of barium and calcium with monocyclic and bicyclic systems also suggests that the cryptate effect is a reflection of a favourable enthalpic term (Anderegg, 1975). Nevertheless, as before, the situation tends to be complicated. For example, nmr and calorimetric studies indicate that the solvation of the sodium and potassium complexes of cryptand 2.2.1 (213; $m = 1$, $n = 0$), as well as of the uncomplexed ligand, varies considerably from solvent to solvent and it is not surprising that such variation has a significant influence on the complexation behaviour of the respective systems (Schmidt, Tremillon, Kintzinger & Popov, 1983).

Further comments. It is apparent from the discussion so far that there are problems in attempting to describe the macrocyclic effect in terms of a set of parameters which will be applicable in every case. Failure to realize this has been a source of the confusion arising from a number of the early studies. Apart from a tendency to argue from the 'specific' to the 'general', other difficulties have arisen from the use of thermodynamic data which is either incomplete for the job in hand or of questionable accuracy. Even when the latter points have been largely satisfied, there is still the uncertainty associated with the choice of a suitable reference system. In any case, as mentioned earlier, detailed (microscopic) analysis of a given macrocyclic effect is only appropriate when the cyclic and non-cyclic systems involved form complexes which have similar coordination geometries. This aspect has tended to be ignored in the past.

Finally, a discussion of the kinetic features of the macrocyclic effect (the 'kinetic macrocyclic effect') mentioned in Chapter 1 is deferred until the next chapter.

6.3   Cyclic ligands and metal-ion selectivity

There has been much interest in macrocyclic ligands for use as metal-ion selective reagents and examples of such selectivity have already been met in Chapter 4. Apart from the usual parameters influencing the metal-ion specificity of open-chain ligands (such as backbone structure and donor atom type), cyclic ligands can be further 'tuned' by variation of the macrocyclic hole size. By this means, a mechanism for discrimination of ions on the basis of their radii may become available. Although studies of this type are of considerable intrinsic interest, they also have implications for a number of areas. These include: aspects of ion storage and transport *in vivo*, the solvent extraction of metals from leach solutions in hydrometallurgy, the synthesis of new chromatography materials for separation of metal ions and the development of metal-ion selective reagents for use in a wide range of analytical, industrial and other applications.

The selective binding of metal ions in solution is thus a well-studied feature of many macrocyclic ligand systems – especially those of the polyether type with non-transition ions. Some examples of this latter type have been discussed already in Chapter 4. A lesser number of studies involving polyaza, polythia, as well as mixed-donor ligand species have been performed, although examples of the selective complexation of a range of transition and other heavy metal ions have also been documented using ligands from these latter categories.

Some general considerations. It is apparent that a close match of the metal-ion radius for the cavity size in a macrocyclic ligand tends to be associated with enhanced stability for the system. Thus, in principle, it is possible to discriminate between closely related metal ions based on their relative fit for the cavity of the ligand. Nevertheless, it needs to be remembered that the selectivity order observed will also be influenced by a range of other factors including the respective solvation patterns for the species involved, the ligand conformations before and after complexation, and the number and nature of the chelate rings formed on complexation. Hole-size effects will tend to be more dominant when the macrocyclic system involved is fairly rigid (such as occurs, for example, in porphyrin ligands); rigid ligands of this type are less able to compensate for a mismatch between the size of the central cavity and the metal-ion involved. For such systems, if the metal ion is smaller than the central cavity size then complex formation will tend to lead to elongated metal-ligand bonds. When the metal ion is too large, either compressed bond-lengths will result or the metal ion will not be positioned in the plane of the donor atom set. Both these situations will be associated with complex destabilization relative to the 'ideal fit' case. It should also be noted that another possibility exists for the 'large metal ion' case: for some transition ions, rather than compressed bond-distances occurring, the metal may change its spin-state with a concomitant change in its covalent radius such that a better fit to the macrocyclic ring occurs. Nevertheless, it needs to be noted that even highly conjugated macrocycles (including the porphyrins) are usually capable of undergoing some limited radial expansion (or contraction) of the central hole.

As the flexibility of the macrocycle increases, then 'mismatch' hole-size effects are expected to be moderated. In any case, as discussed in Chapter 1, a metal ion which is too large for the cavity may be associated with folding of a flexible macrocycle thereby allowing normal metal-ligand bond distances to be achieved. However, this is not always the case, and a number of examples of unfolded macrocyclic complexes containing compressed metal-donor distances are known (Henrick, Tasker & Lindoy, 1985).

Some non-polyether systems. Although the thermodynamics of complexation of the aliphatic N_4-donor macrocycles of type (278; X = NH) (containing 12 to 16 ring members) with metals such as Ni(II) and Cu(II) have been thoroughly investigated, the conformational flexibility of these ligands makes it difficult to define the role of the cavity size in defining the metal-ion specificity patterns observed. Nevertheless, it is clear, for example, that ions such as Cu(II) and high-spin Ni(II) are too large to

X = NH, S

(278)

occupy fully the cavity of the 12-membered ring (that is, they are unable to coordinate such that the metal ion is positioned in the N_4-donor plane). For this series, stability reaches a peak at either the 13- or 14-membered (cyclam) ligand for Cu(II) (Clay, Corr, Micheloni & Paoletti, 1985; Thom & Hancock, 1985) and at the 14-membered ring for Ni(II) (Thom, Hosken & Hancock, 1985). Although the observed selectivity patterns are intimately associated with the magnitude of the respective macrocyclic effects (see earlier discussion), it is difficult to discuss them in a comparative manner since the coordination mode almost certainly changes along each series. Further, the solution structures are in some cases difficult to assign with certainty. An additional complicating factor is present in the case of Ni(II) – an equilibrium between high-spin and low-spin Ni(II) species occurs in aqueous solution for all but the largest ring system.

Detailed studies have been carried out on the interaction of Cu(II) with the 12- to 16-membered tetrathia analogues of type (278; X = S) of the N_4-macrocycles just discussed (Sokol, Ochrymowycz & Rorabacher, 1981; Pett *et al.*, 1983). Once again, although the complex stabilities are markedly ring-size dependent, structural changes occur along the series. The general ease with which Cu(II) is able to accommodate different coordination geometries undoubtedly aids these structural changes. The smaller rings are too small to accommodate Cu(II) and, in the solid state at least, the compounds of the 12- and 13-membered rings have the copper ion 0.53 and 0.38 Å above the respective S_4-donor planes. These complexes have square pyramidal geometries with water molecules occupying the fifth (axial) positions. The 14- to 16-membered ring complexes are tetragonal with coordinated perchlorate groups in axial sites; for these complexes the metal ion lies in the macrocyclic coordination plane. The Cu–S bond lengths increase with increasing ring size, exhibiting average values of 2.30, 2.32, and 2.36 Å for the 14-, 15-, and 16-membered ring structures. Once again, the peak in thermodynamic stability for this series occurs at the 14-membered ring complex.

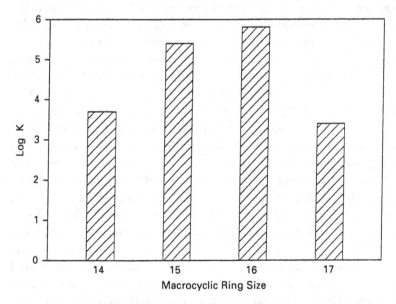

X = 0, S

(279)

An investigation of the hole-size selectivity of the 14- to 17-membered, mixed-donor macrocycles of type (279; X = O) for high-spin Ni(II) has been undertaken. As part of this study, a number of complexes of the type [NiX$_2$(macrocycle)], where X is a halide or pseudohalide ion, were isolated. The X-ray diffraction structures of complexes of the first three (14- to 16-membered) ligands in the series were determined and each of these has a similar pseudo-octahedral geometry in which the macrocycle coordinates in a planar manner with the monodentate ligands occupying axial positions (Henrick *et al.*, 1984). Physical measurements suggested that the remaining complex, containing the 17-membered macrocyclic ring (279; X = O; $n = 4, m = 3$), has a related geometry.

Figure 6.4. The variation of log *K* with macrocyclic ring size for the Ni(II) complexes of O$_2$N$_2$-donor macrocycles of type (279) (Anderegg, Ekstrom, Lindoy & Smith, 1980).

Log K values for each of the nickel complexes were determined in 95% methanol and in water. The effect of ring-size on the respective stabilities is illustrated in Figure 6.4 (Anderegg, Ekstrom, Lindoy & Smith, 1980). A stability maximum occurs for the complex of the 16-membered macrocycle (279; X = O, $n = 3$, $m = 3$). Hole-size calculations based directly on the X-ray data, as well as parallel calculations involving molecular mechanics procedures, strongly suggest that the 16-membered ring provides a cavity which is near-ideal for Ni(II). The stabilities along the series appear to be a direct reflection of the 'fit' of this ion for the available cavities in the respective rings. The presence of dibenzo functions in these macrocycles will reduce their flexibility and thus should help promote hole-size selectivity.

Polyether systems – further considerations. As discussed previously, one of the most widely studied features of macrocyclic and macrobicyclic polyethers is their ability to selectively bind various cations and, in particular, the alkali and alkaline earths. Thus 18-crown-6 forms its most stable complex with potassium for the alkali metals (see Figure 4.5) and with barium for the alkaline earths. In both cases, the ratio of the metal-ion radius to ligand cavity size is approximately unity. However, as already mentioned in Chapter 4, the factors underlying this selectivity are not clear cut. In these cases, the enthalpic origin of the 'peak stability' observed may reflect (in part) enhanced electrostatic interaction when the metal dimensions best match those of the cavity. However, more generally, it is clear that it is difficult to correlate hole size and selectivity for the crown ethers and, for example, in one study it was found that the observed ΔG values for the interaction of 18-crown-6, 15-crown-5 and 12-crown-4 with sodium and potassium were the result of compensating combinations of ΔH and ΔS which arose from a range of thermodynamic causes (Michaux & Reisse, 1982). Further, the importance of other factors besides hole size is underscored by the observation that the normal selectivity of a crown system towards alkali metals can sometimes be altered, and even reversed, as a result of electronic effects arising from substituents on the ring. In any case, the larger-ring crowns are able to wrap around the cation to form a three-dimensional structure in which the concept of a cavity is no longer applicable.

The binding constants are also influenced by the number of ion-dipole interactions present involving the metal and the ether oxygen donors. As the ring size increases, so will the number of available dipoles: this effect will be superimposed (and may dominate) the metal-ion binding pattern along a series of crowns.

Table 6.2. *Thermodynamic data for complexation of alkali metal ions by cryptands in water (Lehn & Sauvage, 1975; Kauffmann, Lehn & Sauvage, 1976).*

Ligand	Parameter	Cation				
		Li$^+$ (0.78)[a]	Na$^+$ (0.98)	K$^+$ (1.33)	Rb$^+$ (1.49)	Cs$^+$ (1.65)
2.1.1 (0.8)[b]	log K	5.5	3.2	<2.0	<2.0	<2.0
	$-\Delta H$[c]	21.3	22.6			
	ΔS[d]	33.5	12.6			
2.2.1 (1.15)	log K	2.5	5.4	4.0	2.6	<2.0
	$-\Delta H$	~0.0	22.4	28.5	22.6	
	ΔS	47.7	25.9	-19.7	-27.2	
2.2.2 (1.4)	log K	<2.0	3.9	5.4	4.4	<2.0
	$-\Delta H$		31.0	47.7	49.4	
	ΔS		-29.3	-59.0	-82.8	
3.2.3 (1.8)	log K	<2.0	1.7	2.2	2.1	2.0
	$-\Delta H$			~12.5	~17.6	22.6
	ΔS			0.0	-19.7	-41.4

[a] Cation radii in Å.
[b] Estimated cavity size (in Å) from models.
[c] ΔH in kJ mol^{-1}.
[d] ΔS in J K^{-1} mol^{-1}.

In general, the cryptands (213) show a stronger correlation between thermodynamic stability and match of the metal ion for the cavity. Thermodynamic data for complexation of the alkali metal ions with a number of cryptands is summarized in Table 6.2. The data for the smaller (less flexible) cryptands 2.1.1, 2.2.1, and 2.2.2 illustrate well the occurrence of 'peak' selectivity.

Calorimetric studies indicate that the enthalpies of complexation tend to show related trends to the observed stability constants and display selectivity peaks, although there is not necessarily a coincidence between the two sets of peaks. Complexation is characterized by the entropy becoming progressively less positive (less favourable) as the cation size decreases. This is illustrated in Figure 6.5 for the complexation of 2.2.1 with the alkali metals.

In contrast to the 'peak selectivity' just discussed, there is evidence that the larger, more flexible, ligands tend to exhibit 'plateau' selectivity – a reflection that a number of the larger metal ions are accommodated by the cryptand without major variation in binding energy.

Metal-ion selectivity by cryptands may be markedly affected by solvent

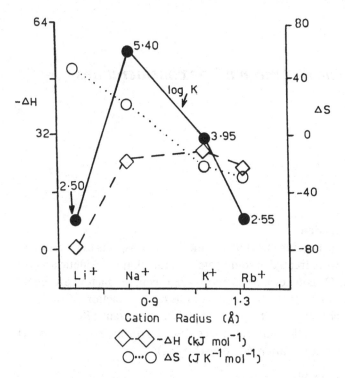

Figure 6.5. Thermodynamic data for complexation of 2.2.1 with the alkali metal ions (in water at 25 °C) (Kauffmann, Lehn & Sauvage, 1976).

change (Abraham, de Namor & Lee, 1977). This is not unexpected since on complexation complete desolvation of the metal ion will normally occur. Thus, the Na^+/K^+ selectivity exhibited by 2.2.1 is strongly solvent dependent (for example, the selectivity is especially enhanced in nitromethane solution) (Schmidt, Tremillon, Kitzinger & Popov, 1983).

In summary, although the metal-ion selectivity of the cryptands is normally largely enthalpy-controlled, entropic terms may also be quite important. Once again, the factors underlying these respective terms may be quite variable and, as a consequence, a criterion for preferred complexation based solely on a match of the cavity for the cation radius may not always be appropriate.

7

Kinetic and mechanistic considerations

7.1   Introduction

In the preceding chapter, thermodynamic aspects of macrocycle complexation were treated in some detail. In this chapter, kinetic aspects are discussed. Of course, kinetic and thermodynamic factors are inter-related. Thus, in terms of a simple complexation reaction of the type given below (charges not shown), the stability constant (K_{ML}) may be expressed directly as the ratio of the second-order formation constant (k_f) to the first-order dissociation rate constant (k_d):

$$M + L \underset{k_d}{\overset{k_f}{\rightleftharpoons}} ML$$

where

$$K_{ML} = \frac{k_f}{k_d}$$

Hence, for reactions of monodentate ligands (or for multidentate ligands in certain cases – see later), K_{ML} can be evaluated solely in terms of the results from kinetic measurements. This has frequently been used as a cross-check of values determined by thermodynamic techniques. Alternatively, K_{ML} values obtained by the latter means have been used in conjunction with either k_f or k_d to obtain the remaining constant.

The kinetics and mechanism of formation and dissociation of macrocyclic complexes is an area covering a wide range of behaviour. Indeed, the mechanistic details of a particular reaction are often closely associated with both the type of metal ion present and the structural features of the cyclic ligand. As such, there are often difficulties in defining general mechanisms which have wide applicability. In this discussion, some representative reactions are considered with emphasis on those features arising from the cyclic nature of the respective systems.

7.2   Formation kinetics

For simple ligand complexes, the formation rate is often control-led by the solvent exchange rate of the solvated metal ion. However, for polydentate (including macrocyclic) ligand systems, the situation is often more complex – especially when the formation of the first ligand-metal bond is not rate-determining. In fact, even if this is not the case, such reactions may be still several orders of magnitude slower than related reactions involving monodentate ligands. A range of factors including the nature of the solvent, steric and electrostatic effects, as well as the possibility of ligand protonation may all influence which step in a multi-step reaction is rate-limiting.

Solvent exchange and complex formation

Where solvent exchange controls the formation kinetics, sub-stitution of a ligand for a solvent molecule in a solvated metal ion has commonly been considered to reflect the two-step process illustrated by [7.1]. A mechanism of this type has been termed a **dissociative interchange** or I_d process. Initially, complexation involves rapid formation of an 'outer-sphere' complex (of ion-ion or ion-dipole nature) which is charac-terized by the equilibrium constant K_{OS}. In some cases, the value of K_{OS} may be determined experimentally; alternatively, it may be estimated from first principles (Margerum, Cayley, Weatherburn & Pagenkopf, 1978). The second step is then the conversion of the outer-sphere complex to an inner-sphere one, the formation of which is controlled by the 'natural' rate of solvent exchange on the metal. Solvent exchange may be defined in terms of its characteristic first-order rate constant, k_{ex}, whose value varies widely from one metal to the next.

$$Ni^{2+} + L \;\underset{\text{Rapid}}{\overset{K_{OS}}{\rightleftharpoons}}\; \underset{\substack{\text{outer}\\\text{sphere}\\\text{complex}}}{Ni^{2+}\ldots L} \;\overset{k_{ex}}{\longrightarrow}\; [NiL]^{2+} \qquad [7.1]$$

For such a mechanism, the overall second-order formation rate con-stant is given by the product of the first-order constant k_{ex} and the equilibrium constant K_{OS}. The characteristic solvent exchange rates are thus often useful for estimating the rates of formation of complexes of simple monodentate ligands but, as mentioned already, in some cases the situation for macrocyclic and other polydentate ligands is not so straightforward.

Polydentate ligand systems

Chelate ring formation may be rate-limiting for polydentate (and especially macrocyclic) ligand complexes. Further, the rates of formation of macrocyclic complexes are sometimes somewhat slower than occur for related open-chain polydentate ligand systems. The additional steric constraints in the cyclic ligand case may restrict the mechanistic pathways available relative to the open-chain case and may even alter the location of the rate-determining step. Indeed, the rate-determining step is not necessarily restricted to the formation of the first or second metal-macrocycle bond but may occur later in the coordination sequence.

In spite of the above, linear aliphatic polyamines such as triethylenetetramine (see L^1 in Table 6.1) appear to react with hydrated Ni(II) such that formation of the first amine-metal bond is rate-determining (Margerum, Rorabacher & Clarke, 1963). The presence of strong donor groups, the relative absence of steric hindrance as well as the flexibility and the favourable relative placement of the donor nitrogens all appear to influence the kinetic behaviour of these systems (Margerum, Cayley, Weatherburn & Pagenkopf, 1978). Conversely, the absence of these properties in a ligand will tend to promote the shift of the rate-determining step to the second (or later) coordination step. This is illustrated by the complexation reactions of corresponding saturated cyclic and non-cyclic polyamines with Cu(II) in basic media (used to eliminate the possibility of ligand protonation causing interference – typically, $Cu(OH)_3^-$ and $Cu(OH)_4^{2-}$ are the reactive species). For complexation of a non-sterically hindered open-chain ligand such as 2.3.2-tet (see 275) with $Cu(OH)_4^{2-}$, first-bond formation appears to remain the rate-determining step whereas, for the cyclam system, the rate-determining step has been assigned to second-bond formation (Lin, Rorabacher, Cayley & Margerum, 1975).

Increased substitution on both cyclic and linear amines is often reflected by a decrease in the corresponding formation rate constants (Drumhiller, Montavon, Lehn & Taylor, 1986). Once again, this is a consequence of the reduced flexibility of the substituted systems since such restrictions will tend to hinder the complexation process. Thus, the reaction of the reduced Curtis ligand with Cu(II) to yield (274) in 0.5 mol dm^{-3} NaOH is 160 times slower than for the analogous reaction with cyclam (Margerum, Cayley, Weatherburn & Pagenkopf, 1978). This difference appears to be a result of the greater resistance of the macrocycle to folding. Unlike open-chain ligands which can encompass the metal ion by means of a 'wrapping' process, coordination of a flexible cyclic ligand is expected to involve

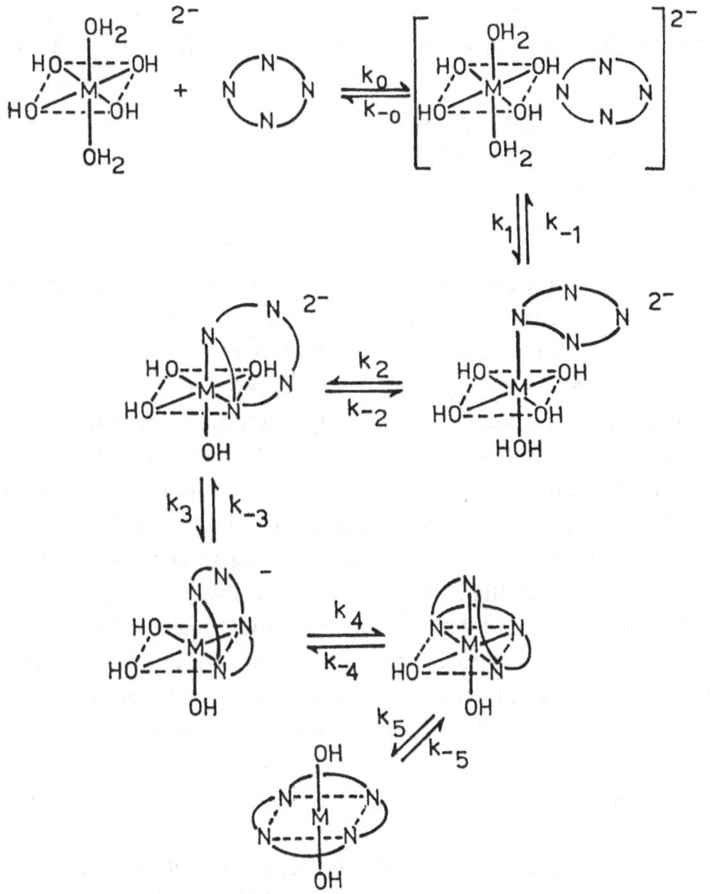

Figure 7.1. The main steps involved in (*trans*) complex formation by a N_4-macrocycle interacting with $[M(OH)_4]^{2-}$.

formation of a folded ligand species during the complexation sequence (see, for example, Figure 7.1). Indeed, it has been possible to isolate such folded intermediates in a number of cases.

During the formation of complexes of highly unsaturated ligands such as the porphyrins (see Chapter 1), ligand folding is unable to occur.

The kinetics and mechanism of metal ion insertion into porphyrins has been the subject of a considerable number of studies. As expected, relative to the formation of complexes of flexible macrocycles, such reactions are slow. For the overall reaction

$$M^{2+} + LH_2 \rightarrow ML + 2H^+$$

(where LH_2 is a neutral porphyrin) the rate equations reported for different divalent metal systems include (Hambright, 1971; Wilkins, 1974):

$$\text{rate} = k'[M^{2+}][LH_2]$$
$$\text{rate} = k''[M^{2+}]^2[LH_2]$$

and

$$\text{rate} = k'''[LH_2].$$

Combinations of these have also been postulated for particular systems. The mechanistic implications of these rate expressions have been considered in detail. However, it is now clear that the observed behaviour for a particular system may be more complex and that these rate expressions may be, at least in some cases, only approximations for more complicated expressions (Shamim & Hambright, 1983; Rao & Krishnan, 1985).

The need for multiple desolvation of the metal ion in some systems may provide a barrier to complex formation which is reflected by lower formation rates – especially for inflexible macrocycles such as the porphyrins. Because of the high energies involved, multiple desolvation will be unlikely to occur before metal-ion insertion occurs; rather, for flexible ligands, solvent loss will follow a stepwise pattern reflecting the successive binding of the donor atoms. However, because of the additional constraints in cyclic systems (relative to open-chain ones), there may be no alternative to simultaneous (multiple) desolvation during the coordination process.

Studies in non-aqueous media. The formation of Ni(II) complexes of both cyclic and open-chain saturated tetraaza ligands has been studied in dipolar aprotic solvents (such as acetonitrile, dimethyl sulphoxide, or dimethylformamide) in order to avoid the ligand protonation which occurs for these strongly basic ligand systems in water (Hay & Norman, 1980; Hertli & Kaden, 1981). For example, in acetonitrile at 25°C, the second-order formation rate constants for cyclam and its substituted derivatives (as well as for three related linear tetraamines) all fall in the range $8(\pm 2) \times 10^2$ dm^3 mol^{-1} s^{-1}; this range lies close to the reported exchange rate for acetonitrile coordinated to Ni(II). Although virtually constant in this solvent, the incorporation rates are nevertheless very dependent on the solvent employed and the evidence suggests that the Eigen–Wilkins I_d mechanism discussed earlier applies in these cases.

Characteristically, for these reactions in anhydrous acetonitrile, at least one additional reaction step is also observed. Whereas the initial solvent exchange step is fast in each case and shows a first-order dependence on the Ni(II) concentration, the subsequent slower reaction is not dependent on the nickel concentration and presumably reflects rearrangement and/or isomerization processes associated with the complex achieving its final coordination geometry. For particular systems there is experimental evidence that inversion of chiral nitrogen centres may be involved in such rearrangements.

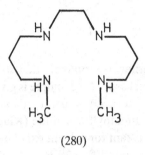

(280)

An investigation of the kinetics of complexation of nickel(II) to cyclam, its open-chain analogue (280), and tetra-*N*-methylated cyclam in dimethyl sulphoxide and dimethylformamide has been performed (Hertli & Kaden, 1981). For a given solvent, the bimolecular rate constants for coordination of cyclam and its open-chain derivative are approximately equal suggesting that a common mechanism applies. Once again, on comparison with the appropriate solvent exchange rate, it is apparent that the rate-determining step reflects the dissociation of the first solvent molecule in the outer-sphere complex. However, the observed slower reactivity of tetra-*N*-methylated cyclam was assumed to be the result of steric influences of the type mentioned previously; in dimethylformamide this ligand reacts 23 times slower than cyclam.

A second slower step is also observed in the case of cyclam and, as before, this is almost certainly a consequence of ligand rearrangement in the coordination sphere of the nickel. Such behaviour is not restricted to reactions in aprotic solvents; analogous two-step complex formation also occurs for the reaction of O_2N_2-donor macrocycles with Ni(II) in anhydrous methanol (Ekstrom *et al.*, 1979). Once again, these systems show a $[M^{2+}]$-dependent incorporation step which is controlled by solvent exchange, followed by a slower nickel-independent rearrangement step. As might be predicted, the rate of this latter step was found to be ring-size dependent.

It is interesting to compare this behaviour with the situation which occurs when a relatively inflexible ligand interacts with a very labile metal ion (that is, one in which the rate of solvent exchange is very high). For

(281)

the reasons discussed earlier, it is unlikely that solvent replacement will be rate-controlling in such cases. An example is given by the formation of the Cu(II) complex of the sterically constrained macrocycle 'tetramethyl-dibenzocyclam' (281) in dimethylformamide (Klaehn, Paulus, Grewe & Elias, 1984). The rate constant for solvent exchange in the case of Cu(II) is quite high, at approximately $10^{8.5}$ s^{-1}. Based on this value, the overall rate of reaction is much slower than expected for the situation where solvent exchange is rate-controlling. In part, this undoubtedly reflects the inherent rigidity of (281): it appears that ligand conformation changes may play a dominant role in the first, copper-dependent step ($k_1 = 39.4$ dm^3 mol^{-1} s^{-1}) of the two steps observed experimentally. The final first-order step ($k_2 = 0.071$ s^{-1}) appears once again to reflect re-arrangement within the coordination shell of the central copper ion.

Effect of ligand protonation. For strongly basic amine ligands, protona-tion will occur spontaneously in water and electrostatic effects between the resulting positively charged ligand and the metal ion may lead to rate decreases relative to the non-protonated cases (Leugger, Hertli & Kaden, 1978).

The formation of the Zn(II), Cd(II) and Pb(II) complexes of the 12- to 15-membered macrocyclic tetraamines of type (278; X = NH) in water has been investigated using a polarographic method (Kodama & Kimura, 1977). The rate law for complex formation in acetate buffers is similar for each system and is given by:

$$\text{rate} = k_1[\text{M}(\text{CH}_3\text{COO})^+][\text{HL}^+] + k_2[\text{M}(\text{CH}_3\text{COO})^+][\text{H}_2\text{L}^{2+}].$$

Hence, under these conditions, both the singularly and doubly-proton-ated ligand species take part in complex formation. Overall, the k_1 values parallel the water exchange rates for the respective hydrated metal ions and thus water exchange appears to be rate-controlling. It should be noted that reaction of the $[\text{H}_2\text{L}]^{2+}$ species with these non-transition ions proceeds too slowly (relative to the mono-protonated ligand species) to be solely a consequence of the additional electrostatic effects present and other factors (as yet unidentified) must also be important.

Polyether complexation. The kinetics of formation of polyether crown, cryptand and related complexes have received considerable attention. Since formation rates are often quite fast, techniques such as temperature-jump, ultrasonic resonance, and nmr have typically been used for such studies.

Complexation between the alkali metal cations and the flexible crowns in methanol approaches the rates expected for methanol exchange in the inner sphere of these cations. The rates are similar to those for the interaction of the natural ionophores such as valinomycin.

It has been proposed that the first stage of alkali metal complex formation by dibenzo-30-crown-10 in methanol involves a fast ligand conformational change which is then followed by a stepwise substitution of the coordinated solvent by the ligand (Chock, 1972):

$$L_1 \overset{\text{fast}}{\rightleftharpoons} L_2$$
$$L_2 + M^+ \underset{k_d}{\overset{k_f}{\rightleftharpoons}} ML_2^+.$$

In this sequence L_1 and L_2 represent different crown conformations.

From ultrasonic absorption experiments, the complexation of K^+ and Cs^+ by 18-crown-6 was proposed to follow the overall two-step mechanism just described (Liesegang, Farrow, Purdie & Eyring, 1976), although corresponding reactions with other monovalent and divalent ions appear to be more complicated (Liesegang *et al.*, 1977; Rodriguez *et al.*, 1977). Indeed, in other studies the Chock mechanism has been rejected in favour of a scheme in which the arrangement of the ligand around the cation is rate-limiting. A mechanism of this type has been assigned to the formation of the Na^+ complex of 18-crown-6 in dimethylformamide (Maynard, Irish, Eyring & Petrucci, 1984).

The dissociation rates for a number of alkali metal cryptates have been obtained in methanol and the values combined with measured stability constants to yield the corresponding formation rates. The latter increase monotonically with increasing cation size (with cryptand selectivity for these ions being reflected entirely in the dissociation rates – see later) (Cox, Schneider & Stroka, 1978).

For non-aqueous solvents, the formation rates for the alkali metal cryptates are not greatly solvent-dependent (Cox, Garcia-Rosas & Schneider, 1981). However, a comparison of the rates for methanol with those for water indicates that the latter are considerably slower (Cox, van Truong & Schneider, 1984) and are, indeed, much slower than expected from the corresponding water exchange rates (Loyola, Pizer & Wilkins, 1977). The reactions in water thus do not correspond to a simple I_d mechanism. For the divalent alkaline earths, the rates in both aqueous and non-aqueous solvents are generally slow and tend to be quite dependent on cation size. Unfortunately, despite a number of studies in the area, the rationale for much of the observed behaviour of these systems remains rather ill-defined.

An investigation of the kinetics of formation of the Li^+ and Ca^{2+} complexes of cryptand 2.1.1 using stopped-flow calorimetry suggests that complexation occurs initially at one face of the cryptand such that the metal is only partially enclosed (to yield an 'exclusive' complex). Then follows rearrangement of this species to yield the more stable product, containing the metal ion inside the cryptand (the 'inclusive' product) (Liesegang, 1981). X-ray diffraction studies have indeed demonstrated that exclusive complexes are able to be isolated for systems in which the metal is too large to readily occupy the cryptand cavity (Lincoln *et al.*, 1986).

7.3   Dissociation kinetics

The dissociation kinetics of macrocyclic complexes have received considerable attention, especially during investigations of the nature of the macrocyclic effect. Before discussing the dissociation of cyclic ligand species, it is of benefit to consider some aspects of the dissociation of open-chain ligand complexes.

The dissociation of metal complexes of polydentate amines is intrinsically a very slow reaction in the absence of acid or other reagent that will aid the process. For example, although the individual steps in the unwrapping of triethylenetetramine (trien) from Ni(II) are far from slow, the respective equilibria corresponding to each step collectively make the process very unfavourable. However, the addition of acid serves to scavenge the amine groups, in a succession of protonation steps, as they dissociate from the central metal. In turn, each amine group is replaced in the coordination shell by a solvent molecule. In the absence of assistance from acid, the dissociation half life for the $[Ni(trien)]^{2+}$ complex is about 16 years (Margerum, Cayley, Weatherburn & Pagenkopf, 1978)! In cases such as this, it needs to be noted that even at neutral pH the hydrogen ion assisted pathway remains important – reflecting the high basicity of the aliphatic amine nitrogens. Other reagents may also assist such dissociation reactions. For example, a second metal ion may be used to scavenge the amines (or other donor groups) as they dissociate; alternatively, the addition of another ligand [such as the cyanide ion or ethylenediaminetetraacetate (edta)] which will bind strongly to the coordination sites as they are vacated may also be effective.

For the situation where the 'natural' rate-determining step lies somewhere along the dissociation sequence (for simple amine polydentate ligands it is expected to be last-bond cleavage or last chelate ring opening), the addition of strong acid may typically force the rate-determining step to occur earlier in the dissociated process. Under such conditions, the production of the protonated species aids removal of the ligand after first-bond cleavage has occurred and, indeed, this cleavage may even become the rate-determining step. Thus, for the dissociation of $[Ni(2,3,2\text{-tet})]^{2+}$ in 0.5 mol dm^{-3} acid, only one step with a rate constant of 0.38 s^{-1} is observed and this was assigned to first-bond cleavage (Vitiello & Billo, 1980). In contrast, for $[Ni(trien)]^{2+}$ the opening of the last chelate ring appears to remain rate-determining; nevertheless, first-order rate constants were also able to be measured for cleavage of the first and second Ni–N bonds for this system (Melson & Wilkins, 1963).

Dissociation of macrocyclic complexes

Amine complexes. As has been mentioned in previous chapters, the dissociation of a macrocyclic ligand from its central metal ion can be extremely slow, even in the presence of strong acid. A classic example of this is provided by the dissociation under acid conditions of the square-planar $[Ni(cyclam)]^{2+}$ species which has been estimated to have a half life of approximately 30 years (Billo, 1984).

Even the complexes of normally labile metals such as Cu(II) may become inert when tightly surrounded by a cyclic ligand and, for example, the red $[Cu(tet\text{-}a)]^{2+}$ isomer (274) shows a dissociation half life in 6.1 mol dm^{-3} HCl at 25°C of 22 days (Cabbiness & Margerum, 1970). Indeed, both the red and the blue isomers of this complex dissociate at rates up to 10^7 times more slowly than the related linear amine complex, $[Cu(2,3,2\text{-tet})]^{2+}$. Such slow dissociation rates are expressions of the kinetic macrocyclic effect. Indeed, much of this inertness is associated with the close fit of these 14-membered tetraaza macrocycles for the ions mentioned. Thus, the Ni(II) and Cu(II) complexes of the related 15-membered ligand containing one additional nitrogen in the macrocyclic ring [namely (282)] are much more labile in acidic solution (Hay, Bembi, Moodie & Norman, 1982). The kinetics of the acid-catalyzed dissociation of these complexes has been investigated. In both cases, a rate law of the type:

$$\text{rate} = k_H[\text{complex}][H^+]^2$$

was observed, indicating the involvement of two protons in the transition

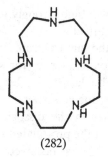

(282)

state of the reaction; for $[CuL]^{2+}$, $k_H = 0.049$ dm^6 mol^{-2} s^{-1} and for $[NiL]^{2+}$, $k_H = 0.63$ dm^6 mol^{-2} s^{-1}.

It is of interest to consider the Cu(II) reaction in more detail. The dissociation of this complex was performed in $HClO_4$ over the concentration range 0.19–0.58 mol dm^{-3}. The reaction was observed to be first-order in complex and a plot of k_{obs} versus $[H^+]$ confirmed the second-order dependence on $[H^+]$. The reaction has been described in terms of the following steps:

$$[CuL]^{2+} + H^+ \overset{K_1}{\rightleftharpoons} [Cu(HL)]^{3+}$$

$$[Cu(HL)]^{3+} + H^+ \overset{K_2}{\rightleftharpoons} [Cu(H_2L)]^{4+}$$

and

$$[Cu(H_2L)]^{4+} \overset{k}{\rightarrow} Cu^{2+}(aq) + H_2L^{2+}$$

It was proposed that mono-protonation of $[CuL]^{2+}$ to yield $[Cu(HL)]^{2+}$ occurs initially at the axial nitrogen atom since this Cu–N bond is expected to be weaker as a result of Jahn–Teller distortion. With respect to this, it should be noted that species such as $[Cu(HL)]^{3+}$ are commonly observed during potentiometric studies of the formation of Cu(II) polyamine complexes. From the proposed mechanism, the following rate law can be deduced:

$$\text{Rate} = \frac{kK_1K_2[CuL^{2+}][H^+]^2}{1 + K_1[H^+] + K_1K_2[H^+]^2}$$

Under the conditions used, $(K_1[H^+] + K_1K_2[H^+]^2) \ll 1$ and if $k_H = kK_1K_2$ the rate $= k_H[CuL^{2+}][H^+]^2$. Details of a possible unwrapping sequence are illustrated in Figure 7.2.

In an extension of this study, the dissociation rates for the copper complexes of the corresponding 16- and 17-membered N_5-macrocycles have been studied (Hay, Bembi, McLaren & Moodie, 1984). Similar rate laws to that just given for the 15-membered ring complex also apply in these cases. It was found that there is an increase in rate as the ring size increases: the respective k_H constants being 0.049 dm^6 mol^{-2} s^{-1} (15-membered), 4.85 dm^6 mol^{-2} s^{-1} (16-membered) and 1.18×10^3 dm^6

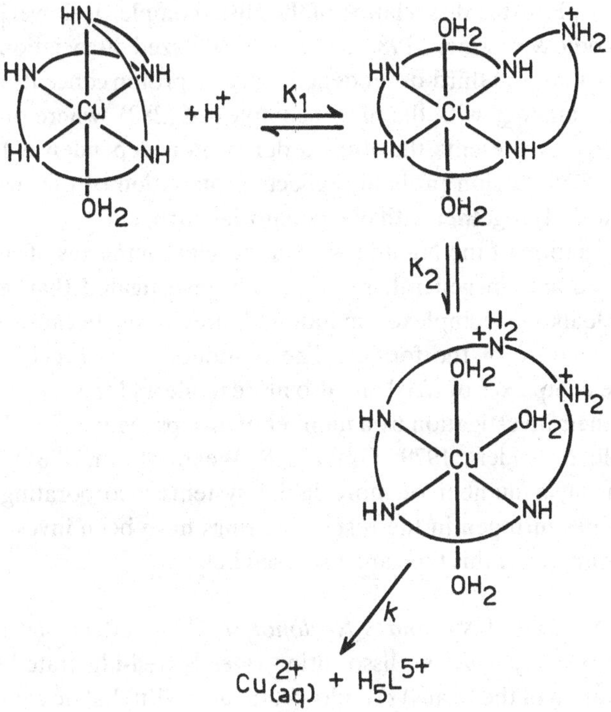

Figure 7.2. The stepwise dissociation of $[CuL]^{2+}$, L = (282), in acid.

$mol^{-2} s^{-1}$ (17-membered). The more facile dissociation in the latter two cases appears to be a direct consequence of the increasing flexibility of the larger rings.

The preparation of the Ni(II) and Cu(II) complexes of the related 18-membered sexadentate ligand (283) has been performed (Hay, Jeragh, Lincoln & Searle, 1978). The Ni(II) complex has a structure of type (284). Once again, both complexes are quite labile under acid conditions with the dissociation of the copper species being relatively

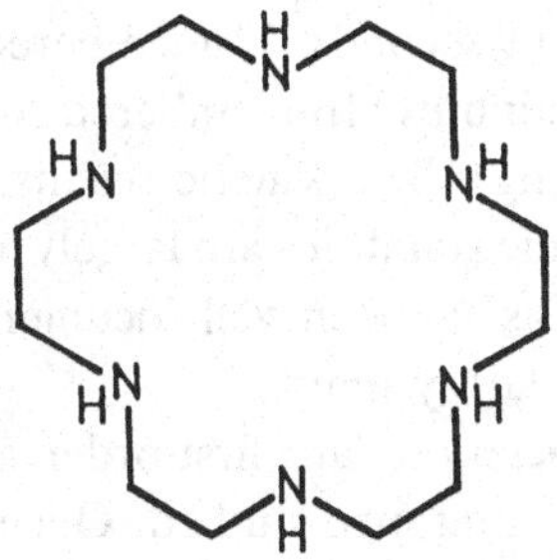

(283)

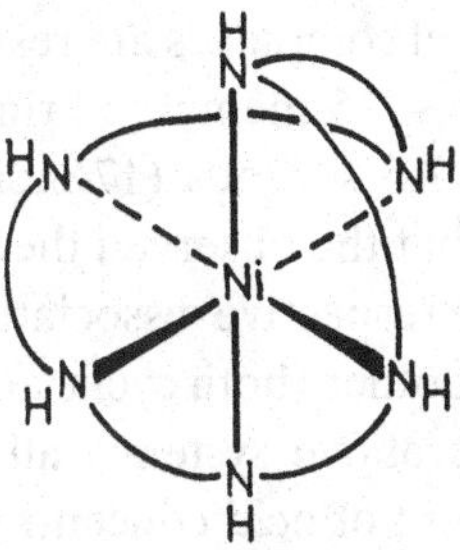

(284)

rapid even at pH 4 [the dissociation of the Ni(II) complex is slower] (Hay, Bembi, Moodie & Norman, 1982). The acid-catalyzed dissociation of the nickel species shows a third-order dependence on proton concentration in this case. By analogy with the nickel complex of (282), where only one axial nitrogen is present, the third-order proton dependence for the system under discussion probably reflects protonation of the two axial nitrogens in (284) together with one equatorial nitrogen.

The dissociation of the N_5- and N_6-macrocyclic complexes of the type just discussed are, in general, more readily investigated than are the tetraazacycloalkane complexes mentioned previously because of the more labile nature of the former. The N_3-donor macrocyclic systems [such as the complexes of (23)] are also more labile and this has similarly facilitated their investigation by a number of groups (Murphy & Zompa, 1979; Riedo & Kaden, 1979; Graham & Weatherburn, 1981). For a similar reason, a number of more labile systems incorporating other donors besides nitrogen in the respective rings have been investigated. Selected examples of this type are discussed below.

Systems containing O_2N_2- and S_2N_2-donor sets. The effect that macro-cyclic ring size may have on dissociation rates is well-illustrated by the acid dissociation of the (trans) pseudo-octahedral Ni(II) halide complexes of the 15- to 17-membered, O_2N_2-donor rings of type (279) (Ekstrom, Lindoy & Smith, 1980). As discussed in Chapter 6, the thermodynamic stabilities of the nickel complexes of this ligand series reach a maximum at the 16-membered ring complex and it is clear from X-ray and other studies that this ring provides a hole size which is near-optimum for this ion (Anderegg, Ekstrom, Lindoy & Smith, 1980).

Although the formation rates of the respective nickel complexes do not vary greatly from one complex to the next, the dissociation rates in acid are very sensitive to ring size with the 16-membered ring complex being the slowest to dissociate in both water and 95% methanol. Thus, the kinetic labilities follow the ring size sequence 14 > 15 > 16 < 17. In hydrochloric acid (1 mol dm^{-3}) at 25°C, the half lives for dissociation of these nickel complexes are respectively 0.9 seconds (14-membered ring), 3.3 minutes (15-membered ring), 18.0 minutes (16-membered ring) and less than 1.3 seconds (17-membered ring). The kinetic results clearly indicate that the observed thermodynamic stabilities are largely a reflec-tion of the respective dissociation rates – as has been well documented for a range of other (both cyclic and non-cyclic) systems.

For the above systems, all the dissociations are first-order and are independent of acid concentration over the range studied. Once again, the acid serves to scavenge the free ligand as it dissociates from the nickel. Alternatively, excess Cu(II) has been employed as the scavenger (Ekstrom *et al.*, 1983) and, as expected, similar first-order dissociation constants to those observed using acid were obtained.

The dissociation kinetics of the nickel chloride complex of the 15-membered, S_2N_2-macrocycle of type (279) has also been investigated in the presence of hydrochloric acid (Lindoy & Smith, 1981). This complex has a similar trans-octahedral structure to that of the O_2N_2-donor systems just discussed. For the sulfur-containing complex, two consecutive (acid-independent) first-order steps were observed, with the second being slower than the first. The data are in accordance with the scheme:

$$[NiL]^{2+} \xrightarrow{k_1} \text{intermediate} \xrightarrow{k_2} \text{products}$$

and there is evidence that the two observed steps reflect successive dissociation of the sulfur donors from the central metal.

S_4-donor systems. The dissociation kinetics of Cu(II) cyclic polythioether complexes have been investigated in a range of methanol/water mixtures (Diaddario *et al.*, 1979). Because of the low basicity of the thioether functions, dissociation will not be accompanied by significant ligand protonation. Once again, these systems are slow to dissociate from the otherwise labile Cu(II) ion. In 80% methanol/water, the dissociation rates do not vary in a regular manner with macrocyclic hole size – probably reflecting the occurrence of coordination geometry changes along the series. The constants for the 12- and 14-membered ring complexes are 4.4 and 9 s^{-1}, respectively; although the 13-, 15-, and 16-membered ring systems dissociate more rapidly, they are still considerably slower than observed for related complexes of open-chain polydentate thioethers.

Crown polyether complexes. The complexation kinetics of a number of crown polyether complexes of alkali metals have been investigated using multinuclear nmr techniques. For example, ^{23}Na nmr has been employed to study the interaction of Na^+ with dibenzo-18-crown-6 in dimethylformamide, dimethoxymethane and methanol (Shchori, Jagur-Grodzinski & Shporer, 1973). The activation energy for the decomplexation reaction is only slightly influenced by the nature of these solvents. Using a similar technique, the interaction of a number of (other) crowns and cryptands with Na^+ has been examined. As in other studies, the cation exchange between free and complexed sites was generally rapid (Lin & Popov, 1981). Nevertheless, the exchange rate is anion-dependent and, for example, for the $NaB(C_6H_5)_4$/18-crown-6/tetrahydrofuran or

$NaB(C_6H_5)_4$/18-crown-6/dioxolane systems the exchange was found to be slow on the nmr timescale: two ^{23}Na resonances were observed for solutions containing an excess of the sodium salt.

Similarly, ^{39}K nmr line-shape analysis has been applied to the study of the interaction of K^+ with 18-crown-6 in a range of solvent mixtures (Schmidt & Popov, 1983). In most instances, the cation exchange proceeds via a normal dissociative process (as found in water); however, in 1,3-dioxolane, metal exchange switches to a bimolecular exchange mechanism.

The few examples discussed so far illustrate the varied behaviour that may be exhibited by alkali metal/crown ether systems. Unfortunately, as yet, the reason for much of this variation is little understood.

In water, the relatively low stability of the alkali metal and alkaline earth cryptates (except those for which there is a near-optimal fit of the cation in the intramolecular cavity) has resulted in difficulties in undertaking a wide-ranging kinetic study in this solvent. However, in non-aqueous media, the stability constants are larger and most of the studies have been performed in such media.

The kinetics and thermodynamics of complex formation in methanol for the interaction of cryptands 2.1.1, 2.2.1 and 2.2.2 with the alkali metal

Table 7.1. *Formation* (k_f) *and dissociation* (k_d) *rate constants for the alkali metal complexes of the cryptands of type* (213) *in methanol at 25 °C (Cox, Schneider & Stroka, 1978).*

Cryptand	Cation	$\dfrac{k_d}{s^{-1}}$	$\dfrac{k_f{}^a}{dm^3\ mol^{-1}}$	$\dfrac{k_{H^+}{}^b}{dm^3\ mol^{-1}}$
2.1.1	Li^+	4.4×10^{-3}	4.8×10^5	4.9×10^{-1}
2.1.1	Na^+	2.50	3.1×10^6	0
2.2.1	Li^+	7.5×10	1.8×10^7	2.1×10^3
2.2.1	Na^+	2.35×10^{-2}	1.7×10^8	3.7×10^{-1}
2.2.1	K^+	1.09	3.8×10^8	—
2.2.1	Rb^+	7.5×10	4.1×10^8	—
2.2.1	Cs^+	$\sim 2.3 \times 10^4$	$\sim 5 \times 10^8$	
2.2.2	Li^+	$> 3 \times 10^2$		
2.2.2	Na^+	2.87	2.7×10^8	4.2×10^2
2.2.2	K^+	1.8×10^{-2}	4.7×10^8	0
2.2.2	Rb^+	8.0×10^{-1}	7.6×10^8	0
2.2.2	Cs^+	$\sim 4 \times 10^4$	$\sim 9 \times 10^8$	

[a] k_f obtained from $k_f = k_d K$.
[b] $k_{H^+} = k_{obs} - k_d/[H^+]$.

cations indicate that the dissociation rates in this solvent are slower than the corresponding rates in water (Cox, Schneider & Stroka, 1978). The dissociation rates were obtained from the reaction of the complexes with acid (the formation rates were then obtained from the dissociation rates via the measured stability constants). As mentioned earlier in this chapter, the selectivity of the cryptands for alkali metal cations was found to be reflected solely in the respective dissociation rates. A summary of the rate constants for complex formation and dissociation for this series of ligands is given in Table 7.1. (Cox, Schneider & Stroka, 1978).

As well as an acid-independent pathway, an acid-dependent pathway also occurs for the dissociation of a number of the alkali and alkaline earth metal cryptates (Table 7.1). Similar behaviour to that just discussed for methanol also occurs for reactions in a range of other non-aqueous solvents (Cox, Truong & Schneider, 1984).

The kinetics of formation and dissociation of the Ca^{2+}, Sr^{2+} and Ba^{2+} complexes of the mono- and di-benzo-substituted forms of 2.2.2, namely (214) and (285), have been studied in water (Bemtgen *et al.*, 1984). The introduction of the benzene rings causes a progressive drop in the formation rates; the dissociation rate for the Ca^{2+} complex remains almost constant while those for the Sr^{2+} and Ba^{2+} complexes increase. All complexes undergo first-order, proton-catalyzed dissociation with $k_{obs} = k_d + k_H[H^+]$. The relative degree of acid catalysis increases in the order $Ba^{2+} < Sr^{2+} < Ca^{2+}$ for a given ligand. The ability of the cryptate to achieve a conformation which is accessible to proton attack appears to be inversely proportional to the size of the complexed metal cation in these cases.

(285)

Metal exchange reactions

For complexes of open-chain polydentate ligands, metal exchange usually shows a dependence on the concentration of the exchanging metal ion (Wilkins, 1974) and generally such reactions proceed via formation of a binuclear complex. However, with cyclic ligands, bridge formation may be more difficult and spontaneous dissociation of the complex may dominate the kinetic behaviour. Thus, there was no observed dependence on the copper concentration when this ion was exchanged for nickel in an S_2N_2-donor aliphatic macrocycle (Kallianou & Kaden, 1979). However, metal exchange involving porphyrin systems shows in some cases a dependence on both the macrocyclic complex and the exchanging metal ion (Grant & Hambright, 1969); in such cases the metal exchange is slow and, once again, it is assumed that a binuclear activated complex is involved.

7.4   The kinetic macrocyclic effect – final comments

It is now appropriate to summarize the main features of the kinetic macrocyclic effect. Relative to the corresponding open-chain complexes, slow kinetics of dissociation is a characteristic of a large number of the macrocyclic systems studied. This behaviour reflects the fact that stepwise removal of a cyclic ligand from the coordination sphere of a metal tends to be more difficult than for an open-chain analogue largely because the cyclic ligand has no end. Indeed, the coordinated cyclic ligand may require major (unfavourable) rearrangement, such as folding, within the coordination sphere before dissociation can occur.

Although the rates of formation may sometimes be somewhat slower in the case of macrocyclic complexes, a greater decrease normally occurs for the dissociation rates; it is the latter which is reflected in the enhanced thermodynamic stabilities ($K = k_f/k_d$) which are characteristic of the macrocyclic effect. Even though these properties reflect the cyclic nature of the macrocycle involved, it is worth noting that similar effects may occur for non-cyclic systems in which the donor atoms are constrained by attachment to a rigid backbone. That is, it has become increasingly apparent that many of the properties associated with simple macrocyclic systems also occur for metal ions bound in structurally rigid sites which are found, for example, in a range of metalloenzymes.

8

Redox properties

8.1 Introduction

The redox chemistry of macrocyclic ligand complexes has received much attention. There are several reasons for this.

(i) The inertness of many macrocyclic systems makes them attractive for electrochemical studies since the redox changes are less likely to be influenced by competing equilibria involving ligand dissociation than non-cyclic systems.

(ii) Macrocyclic ligand systems tend to provide a well-defined environment for the metal ion which, in the case of the more rigid ligands, will not vary greatly from reactant to product. Such geometric constraints will influence the electron transfer kinetics of a particular redox couple as well as the thermodynamics of the system. When the macrocyclic cavity is close to that preferred by the reduced species then a high standard potential will occur whereas if it is more favourable to the oxidized state then a lower standard potential ($E^{\ominus}$) will be the result. Cyclic systems thus provide an additional parameter, the macrocyclic cavity size, which can be used to 'tune' the potential of a particular redox reaction.

(iii) Many of the synthetic macrocycles may be considered to be models for natural cyclic systems; a number of the latter are central to *in vivo* redox behaviour.

(iv) There is an interest in obtaining stable, water-soluble, redox reagents based on inexpensive materials for use in such devices as photoelectrochemical cells and redox storage batteries (Chen & Bard, 1984). Macrocyclic systems appear to be candidates for such applications.

(v) Redox reactions on cyclic complexes have frequently yielded products in which the unsaturation pattern of the starting (complexed) macrocycle has been altered. Commonly, the central metal plays both a catalytic and directing role in such behaviour. Reactions of this type, as mentioned in Chapter 2, have been used synthetically to obtain complexes exhibiting a required unsaturated pattern, or fully saturated pattern, as the case may be.

In this chapter, for convenience of discussion, the redox behaviour of cyclic systems is divided into two categories. Initially, those reactions for which there is no marked alteration of the unsaturation pattern of the ligand are discussed. Subsequently, systems in which the *final product* involves an alteration in the ligand's unsaturation are treated. However, it is emphasized that a continuum exists between redox behaviour which is completely metal-centred and that which is solely ligand-based.

8.2   Stabilization of less-common oxidation states
General considerations

The capacity of cyclic ligands to stabilize less-common oxidation states of a coordinated metal ion has been well-documented. For example, both the high-spin and low-spin Ni(II) complexes of cyclam are oxidized more readily to Ni(III) species than are corresponding open-chain complexes. Chemical, electrochemical, pulse radiolysis and flash photolysis techniques have all been used to effect redox changes in particular complexes (Haines & McAuley, 1982); however the major emphasis has been given to electrochemical studies.

It needs to be pointed out that $E^{\ominus}$ values may also be quite sensitive to the nature of the solvent and supporting electrolyte used for an electrochemical study. Apart from solvation effects of the non-specific type, solvent molecules may occupy coordination sites in either the starting complex or the products and hence influence redox behaviour (Fabbrizzi, 1985). Similarly, the nature of the anion present may also strongly influence the redox potential if it has ligating properties (Zeigerson *et al.*, 1982). Because of such effects, caution needs to be exercised in attempting to compare electrochemical data which have not been obtained under similar conditions.

In the promotion of less-common oxidation states, much attention has been focused on the redox behaviour of transition metal ions such as nickel and copper although many other metal types have also had unusual oxidation states stabilized by macrocyclic ligands. However, within the limitations of a single chapter it is not possible to attempt a wide ranging discussion of the area. Instead, a somewhat limited number of studies will be described with emphasis being given to selected nickel and copper systems.

Representative oxidations

Ni(III) and Cu(III) complexes. In early classic studies the redox chemistry of tetraaza macrocyclic complexes of Ni(II) and Cu(II) (of the Curtis and reduced Curtis type) was investigated in acetonitrile (Olson & Vasilevskis, 1969; 1971). These authors were the first to report the electrochemical generation of Ni(III) and Cu(III) complexes of such N_4-cyclic ligands. Since this time, a considerable number of related studies, involving both nickel and copper macrocyclic species, have been reported.

In 1974, an investigation of the electrochemical behaviour of twenty-seven low-spin Ni(II) complexes of tetraaza macrocycles was carried out in which the factors influencing the generation of both Ni(I) and Ni(III) species were investigated (Lovecchio, Gore & Busch, 1974). The degree of stabilization for each of these oxidation states was, as expected, found to be very dependent on the structure of the macrocycle involved. In particular, it was demonstrated that the overall redox properties of a given system are influenced by macrocyclic ring size, the charge on the ligand, the nature of ligand substituents and the extent and type of ligand unsaturation. An empirical partitioning of the electronic and structural factors controlling redox behaviour was found to be possible. To a lesser extent, similar partitioning of related systems involving Fe(II)/Fe(III), Co(II)/Co(III) and Cu(II)/Cu(III) couples was also possible (Busch, 1978).

The $E_{1/2}$ values corresponding to the Ni(II)/Ni(III) couple for a selection of tetraaza macrocyclic complexes in acetonitrile are given in Figure 8.1. Overall, for such square-planar Ni(II) complexes, a potential range spanning almost two volts has been observed.

It needs to be stressed that there is not necessarily a direct correlation between ring size and $E_{1/2}$ values for the complexes of a related series of cyclic ligands (Fabbrizzi, 1979). Thus, for the series of 13- to 15-membered aliphatic N_4-macrocycles of type (278; X = NH), it is the 14-membered cyclam ring which most favours the oxidation of Ni(II) to Ni(III) (Fabbrizzi, 1985). In cases such as this, the degree of ring strain (as well as ring size) undoubtedly influences the redox behaviour. Further, it has been proposed that the stronger the in-plane ligand field the more readily oxidation to the trivalent state occurs. Such behaviour appears to reflect the raising of the energy of the $d_{x^2-y^2}$ orbital (the 'redox orbital') by strong equatorial interactions, thus allowing easier removal of an electron from

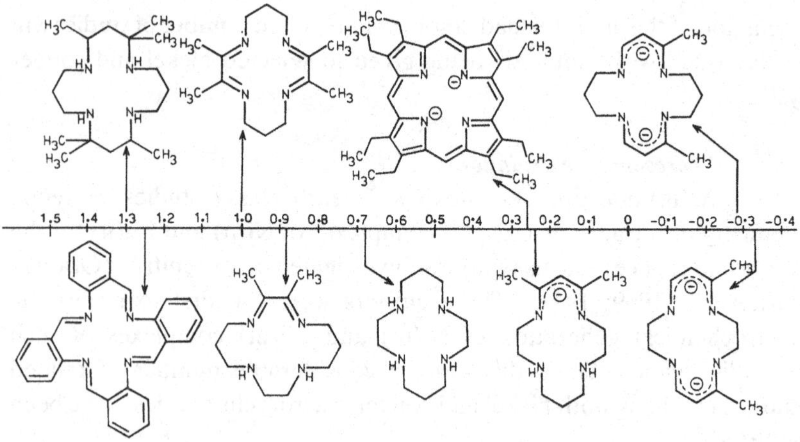

Figure 8.1. $E_{1/2}$ values for the Ni(II)/Ni(III) couple of N_4-donor macrocyclic systems in acetonitrile (versus Ag/AgNO$_3$).

the Ni(II) precursor. For the series just discussed, it appears that cyclam interacts (equatorially) most strongly with the Ni(II) ion. Certainly, if isocyclam (286) is substituted for cyclam then the oxidation becomes more difficult; relative to cyclam, the donor atoms of (286) are not so ideally positioned for coordination to nickel.

In order to investigate further the relationship between in-plane interaction and ease of oxidation of the central metal ion, a study of the oxidation of Ni(II) and Cu(II) complexes of the cyclam derivative (287) was investigated (Fabbrizzi, 1985). This 'dioxocyclam' ligand (287) coordinates to a divalent metal ion with deprotonation of the amido groups.

In oxidation reactions starting from the Ni(II) or Cu(II) complexes, the ligand's dinegative charge on coordination should aid stabilization of the tripositive charge on the oxidized metal ions. Thus, it was found that both complexes undergo reversible one-electron oxidations which occur more readily (at less positive potentials) than for the corresponding cyclam systems.

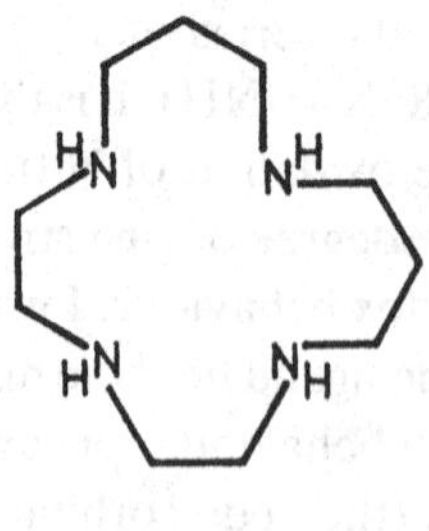

(286) 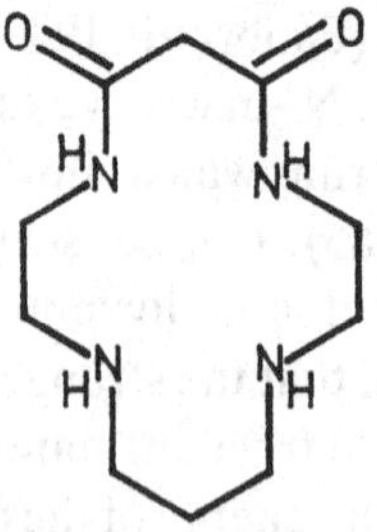   (287)

(n = 2 or 3)

(288)

As an extension of the study involving the complexes of dioxocyclam, a parallel investigation has been performed on the related binuclear derivatives of type (288) (Buttafava *et al.*, 1984). This neutral di-Cu(II) species undergoes reversible oxidation to the corresponding di-Cu(III) species according to two one-electron steps separated by only 100 mV. Even so, this difference is larger than expected on statistical grounds (36 mV) and reflects a small but detectable mutual interaction between the two redox centres. As a result of the very small difference in potential between the pairs of copper ions in this binuclear complex, the latter may prove useful as a practical 'two-electron' redox reagent.

It needs to be noted that when the ligand system contains extensive unsaturation, then oxidation of the corresponding complex may yield a product containing a stabilized cation radical (rather than one in which the metal oxidation state has been altered). For example, such a situation has a tendency to occur on oxidation of divalent metal complexes [including Ni(II)] of the tetraphenyl-substituted porphyrin macrocycle.

Selected other metal ion systems. Largely because of the possible involvement of manganese in the photosynthetic production of oxygen, the redox behaviour of a range of manganese species has been investigated – especially those containing highly-conjugated macrocycles such as the porphyrins and phthalocyanines (see Chapter 1). The Mn(II) phthalocyanine complex readily undergoes a one-electron oxidation to yield the corresponding (formally) Mn(III) species. Further oxidation of the latter is possible but such subsequent reactions appear to be essentially ligand-centred (Lever, Minor & Wilshire, 1981).

The porphyrin complexes of manganese exist in a range of formal oxidation states of which the +3 state is most stable. Mn(III) porphyrins may be electrochemically oxidized in non-aqueous media to yield products which are probably Mn(III) cation radicals or di-cations (Kelly & Kadish, 1982). The potentials at which these reactions occur are sensitive to the basicity of the porphyrin ring, the solvent type, the counter-ion used and the nature of any axially-bound ligands. The chemical oxidation of Mn(III) porphyrins has also been widely investigated using a variety of oxidants. Water-soluble Mn(III) porphyrins undergo oxidation in aqueous alkaline solution to yield 'Mn(IV) porphyrins' which may exist as μ-oxodimers (Carnieri, Harriman, Porter & Kalyanasundaram, 1982). When hypochlorite is the oxidant a further oxidation step occurs to yield a species which has been postulated to be a Mn(V) oxo-porphyrin. Such oxidized products are of limited stability and revert to Mn(III) species on standing, even in the dark.

In an important development, it has been demonstrated that Mn-porphyrins in the presence of a variety of reduced oxygen-containing species (such as iodosylbenzene) act as catalysts for the oxidation of a number of organic molecules including such normally unreactive compounds as the alkanes (Hill & Schardt, 1980). Highly oxidized manganese species have been implicated in these synthetically promising reactions. Related activated (oxo-) iron porphyrins also exhibit similar behaviour (Groves & Gilbert, 1986) and in this regard mimic the reactivity of cytochrome P-450 (see Chapter 9).

In another study, Co(III) has been shown to be favoured over Co(II) as the ligand strain energy decreases in the complexes of the N_4-donor aliphatic macrocycles (278; X = NH). Thus, for the 13- to 16-membered ring complexes of type *trans*-$[CoCl_2L]^{2+}$, it is the complex of cyclam (14-membered) which most promotes the Co(II)/Co(III) transformation (Hung *et al.*, 1977).

In the presence of a saturated tetraaza macrocycle such as cyclam, disproportionation of Ag(I) occurs to produce a silver mirror and a stable Ag(II) complex of the macrocycle (Kestner & Allred, 1972; Barefield & Mocella, 1973). In some cases the Ag(II) complexes so formed may then be oxidized further to Ag(III) species either electrochemically or chemically [using nitrosyl (NO^+) salts].

Numerous other examples of related oxidations involving a variety of macrocyclic types and metal ions have been reported. However, the examples given so far serve to illustrate the main features of such reactions. It is appropriate that we now give consideration to some representative reductions of macrocyclic species.

Representative reductions

Macrocyclic ligands have also been effective in stabilizing less-common, lower oxidation states for a variety of metal ions. In particular, a considerable number of investigations have been concerned with the formation of stable Ni(I), Cu(I), Fe(I) and Co(I) species.

Ni(I) *and Cu*(I) *complexes.* For nickel, it has been generally observed that the lower the oxidation potential for the Ni(II)/Ni(III) couple, the more negative is the reduction potential for the Ni(II)/Ni(I) couple.

Esr spectroscopy has proved useful for characterizing the reduced species, enabling assessment of whether the unpaired electron density is localized on the metal or delocalized onto the ligand. A typical study is the reported electrochemical reduction of the Ni(II) complex of (289) (Bailey, Bereman, Rillema & Nowak, 1984); a reversible one-electron reduction occurs to yield a product whose esr spectrum shows two g values which are characteristic of a Ni(I) derivative. In contrast, the reduction product formally represented by (290) has been shown to have extensive delocalization onto the ligand; it is probably best described as involving coordination of a free radical to a central Ni(II) (Lovecchio, Gore & Busch, 1974).

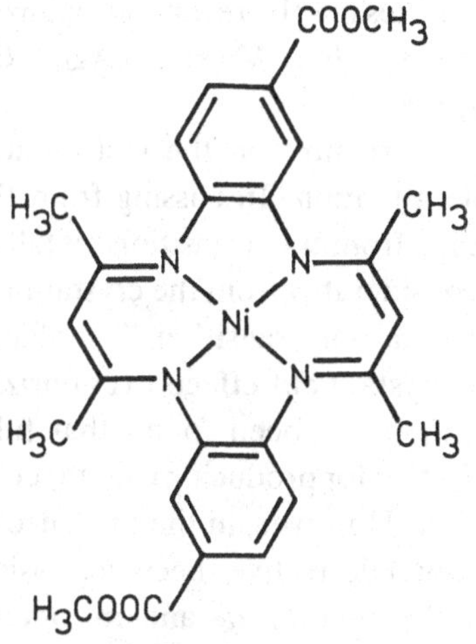

(289)

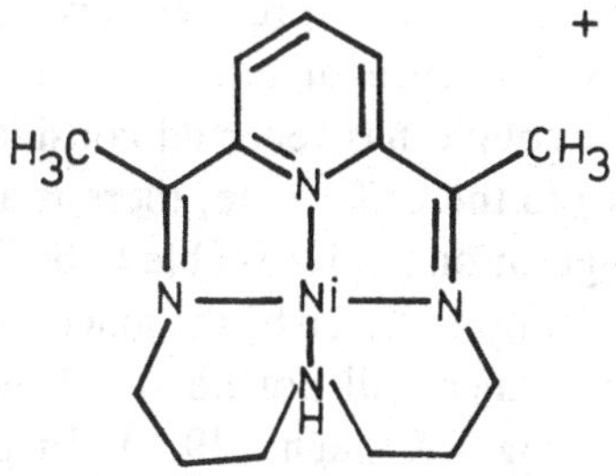

(290)

The electrochemistry of a range of Ni(II) porphyrins and chlorins has been investigated. All complexes are reduced by a similar one-electron mechanism which appears to involve the formation of anion radicals (Chang, Malinski, Ulman & Kadish, 1984).

Stable Cu(I) complexes of tetraaza macrocycles are able to be generated in oxygen-free aprotic solvents (Palmer, Papaconstantinou & Endicott, 1969; Olson & Vasilevskis, 1971). In aqueous solvents there is a tendency for such species to decompose via loss of ligand (Freiberg, Meyerstein & Yamamoto, 1982). Indeed, aqueous Cu(I) is unstable with respect to disproportionation to Cu(II) and elemental copper. However, extensive N- and C-alkylation of the macrocycle, as occurs in (291), slows

(291)

the rate of ligand exchange so that the Cu(I) species becomes stable even in air-saturated aqueous solutions (this species may be reversibly oxidized to the corresponding divalent complex at +0.47 V versus Ag/AgCl) (Jubran, Cohen, Koresh & Meyerstein, 1984).

The influence of tetraaza macrocycle structure on the Cu(II)/Cu(I) redox couple has received considerable attention. In passing from the Cu(II) to the Cu(I) state, there is a change from the 'transition metal' d^9 configuration (which is able to benefit considerably from the crystal-field stabilization offered by the macrocycle) to the non-transition d^{10} configuration (which will not be stabilized by crystal-field effects) (Fabbrizzi, Lari, Poggi & Seghi, 1982). In general, it has been found that fully saturated tetraaza macrocycles are ineffective for producing Cu(I) species stable enough to resist disproportionation. However, in the presence of unsaturation, the generation of stable Cu(I) derivatives becomes easier. This effect appears to be related to the relative σ- and π-bonding characteristics of the Cu(II) and Cu(I) complexes involved. The evidence suggests that the Cu(II)/Cu(I) transformation in macrocyclic systems may be promoted by decreasing the σ-donor ability of the macrocycle through ligand modification [to destabilize Cu(II)] or, alternatively, by increasing its π-acceptor properties [thereby stabilizing Cu(I)].

It has been recognized that sulfur donors aid the stabilization of Cu(I) in aqueous solution (Patterson & Holm, 1975). In a substantial study, the Cu(II)/Cu(I) potentials and self-exchange electron transfer rate constants have been investigated for a number of copper complexes of cyclic poly-thioether ligands (Rorabacher et al., 1983). In all cases, these macrocycles produced the expected stabilization of the Cu(I) ion in aqueous solution. For a range of macrocyclic S_4-donor complexes of type

(278; X = S) and corresponding open-chain complexes, the self-exchange rate constants were found to be of similar magnitude (with an overall variation of less than seven-fold). This result is perhaps somewhat unexpected in view of the considerable differences in rearrangement (between the cyclic and open-chain species) which might be expected to accompany electron transfer in these systems.

Selected other metal ion systems. There have been a number of investigations of the reduction of iron macrocyclic ligand complexes. In one such study, the Fe(II) complex $[FeL(CH_3CN)_2]^{2+}$ [where L = (292)] was shown to exhibit three reduction waves in acetonitrile (Rakowski & Busch, 1973). Controlled-potential electrolysis at the first reduction plateau (-1.2 V) led to isolation of $[FeL]^+$ for which the esr spectrum is typical of a low-spin Fe(I) system. The quasi-reversible Fe(I)/Fe(II) couple occurs at -0.69 V versus $Ag/AgNO_3$.

Co(I) species of a wide range of N_4-macrocycles may be generated in a related manner. The stabilization of this oxidation state is perhaps not surprising since the macrocycles used readily adopt a square-planar geometry which also commonly stabilizes other diamagnetic d^8 systems. The Co(II)/Co(I) couple for the complexes of a number of tetraaza macrocycles has been demonstrated to vary in a predictable manner as the degree of ligand unsaturation alters (Tait, Lovecchio & Busch, 1977). In contrast, there is no clear evidence that the fully saturated N_6-cage ligands (see Chapter 3) stabilize Co(I) (Bond, Lawrance, Lay & Sargeson, 1983). These macrobicyclic cages constrain the cobalt ion in a six-coordinate environment such that attainment of a favourable square-planar geometry on reduction is not possible.

(292)

Studies involving latter-row transition complexes of both macrocyclic and macrobicyclic ligands have been performed. For example, the reduction of Ru(III) complexes of type $[RuLCl_2]^+$ [where L represents the 14- to 16-membered, aliphatic N_4-macrocycles of type (278; X = NH)] has been studied (Walker & Taube, 1981). For these, the Ru(III)/Ru(II) couple is not appreciably affected by an increase in the ring size of the macrocycle (a variation of only 55 mV occurs from the 14- to the 16-membered ring complex); but the potential of the couple does become slightly more positive, reflecting an increased stability of the Ru(II) complex relative to the Ru(III) species. This is the expected trend, since the larger Ru(II) ion will be more readily accommodated as the macrocyclic ring size increases. The small magnitude of the effect has been ascribed to the electron involved in the reduction entering a non-bonding π-d orbital whose energy is little affected by the nature of the macrocyclic ligand present.

The kinetically-stabilized complexes of the cage ligands normally yield redox reagents free of the exchange problems often associated with simple complexes. Indeed, the redox chemistry of the complexes shows a number of unusual features; for example, saturated cages of the type mentioned in Chapter 3 are able to stabilize rare (monomeric) octahedral Rh(II) species (d^7 electronic configuration) (Harrowfield *et al.*, 1983). In a further study, radiolytical or electrochemical reduction of the Pt(IV) complexes of particular cages has been demonstrated to yield transient complexes of platinum in the unusual 3+ oxidation state (Boucher *et al.*, 1983).

Various manganese complexes have also been investigated. In one study, six-coordinate compounds of type $[MnXL]^+$ and seven-coordinate species of type $[MnX_2L]$ [where L = (293) and X is one of a range of

(293)

monodentate ligands] were used as the starting complexes for cyclic voltammetric investigations (Ansell, Lewis, Ramsden & Schroder, 1983). Since (293) is potentially a π-acceptor macrocycle, it was of interest to discover whether Mn(I) or even Mn(0) species might be stabilized. However, the studies indicated that a reversible one-electron reduction wave occurs near -1.4 V (versus Ag/AgNO$_3$) in each case. The observed reduction potential showed little variation with change of axial ligand and this suggested that the redox behaviour is largely associated with the respective coordinated macrocycles rather than with the metal centre. It proved possible to isolate the corresponding one-electron reduction products after using controlled-potential electrolysis to quantitatively reduce the starting complexes. Esr confirmed that these products are Mn(II) ligand-radical species and it was postulated that the added electron is delocalized in the conjugated N$_3$-portion of the ligand. In a further attempt to obtain Mn(I) species, π-acceptor ligands such as CO and phosphines were added to the solution in the electrochemical cell. However, once again, no metal-ion reduction was detected on electrolysis of the solution.

The study just described is in accordance with the observation that electrochemical reduction of the (highly conjugated) phthalocyanine (5) complex of Mn(II) also gives no evidence for the formation of a Mn(I) species (in contrast to the corresponding iron and cobalt complexes which, on reduction, yield Fe(I) and Co(I) products) (Lever, Minor & Wilshire, 1981).

8.3   Reactions involving change of ligand unsaturation

As mentioned previously, a large number of redox reactions involving macrocyclic ligand complexes have resulted in discrete changes in the unsaturation pattern of a variety of macrocyclic systems. Chemical, electrochemical, and catalytic reactions have been widely used to change the level of unsaturation in such systems. Although the mechanisms of the majority of such transformations are not well understood, it is clear that the reactions tend to proceed via prior change in the oxidation state of the central metal ion.

Oxidative dehydrogenations of many macrocyclic ligand complexes have now been documented. Typically, these reactions involve conversion of coordinated secondary amines to imine groups.

In early studies, reaction of the Ni(II) complexes (59) and (60) of the *trans* and *cis* (diimine) isomers of the Curtis macrocycle with nitric acid yielded the tetraimine species (294) and (295), respectively. There is strong evidence that these reactions proceed via Ni(III) intermediates

(294)   (295)

(Curtis, 1968; 1971; Curtis & Cook, 1967). In accordance with this, an analogous Co(III) complex is not oxidized by nitric acid; there is no readily accessible higher oxidation state of the metal ion in this case.

The reverse reaction, namely hydrogenation, has also frequently been used to decrease the degree of unsaturation present in macrocyclic systems – typically converting imine linkages to amine groups. Such hydrogenations have usually been performed catalytically (for example, using H_2 in the presence of Raney nickel or a precious metal catalyst) or by means of chemical reductants such as sodium borohydride.

By using combinations of hydrogenation and dehydrogenation reactions it has been possible to obtain nickel derivatives of the Curtis macrocycle containing from zero to four imine groups (Curtis, 1968; 1974). Related reactions in the presence of a variety of other central metal ions have been described. The electrochemical oxidation of the Cu(II) complex of the reduced Curtis ligand proceeds initially via a two-electron step to yield the monoimine complex (296) (Olson & Vasilevskis, 1971).

(296)

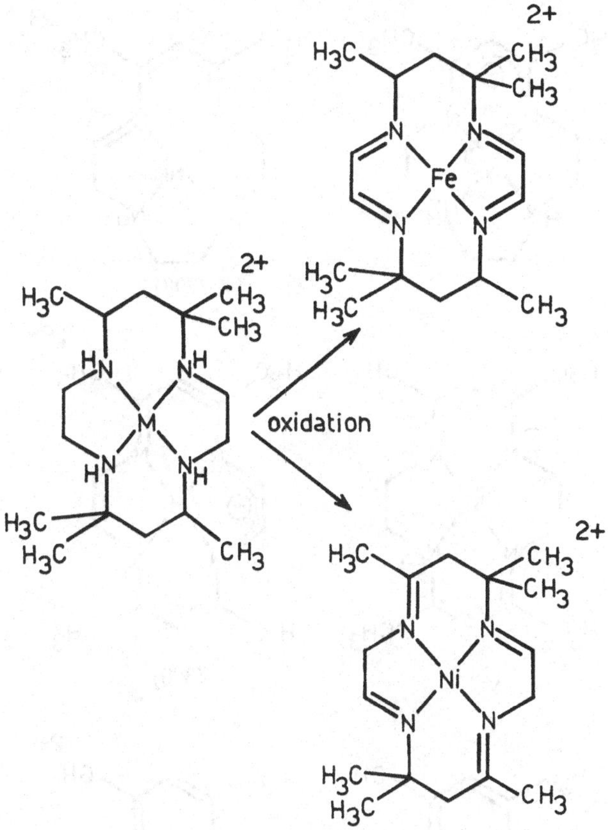

Figure 8.2. The metal-ion dependence of ligand oxidation in complexes of the reduced Curtis macrocycle.

Once again the data implicate a Cu(III) intermediate. In this case, further oxidations subsequently occur to yield a product containing four imine groups.

The location of the induced unsaturation in the macrocyclic system is metal-ion dependent. This is illustrated by the examples given in Figure 8.2. In the Fe(II) complex, the imine functions form as conjugated pairs (Dabrowiak, Lovecchio, Goedken & Busch, 1972; Goedken & Busch, 1972) – such α-diimine species have long been known to have a special affinity for Fe(II). In contrast, Ni(II) promotes formation of a product in which the respective imine functions are in electronically isolated positions (Curtis, 1968; 1974).

A variety of redox reagents have been employed to effect both oxidation and reduction of coordinated macrocycles. For example, the

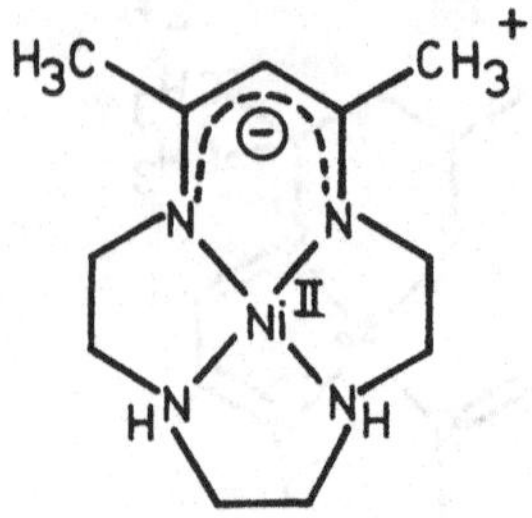

(297)

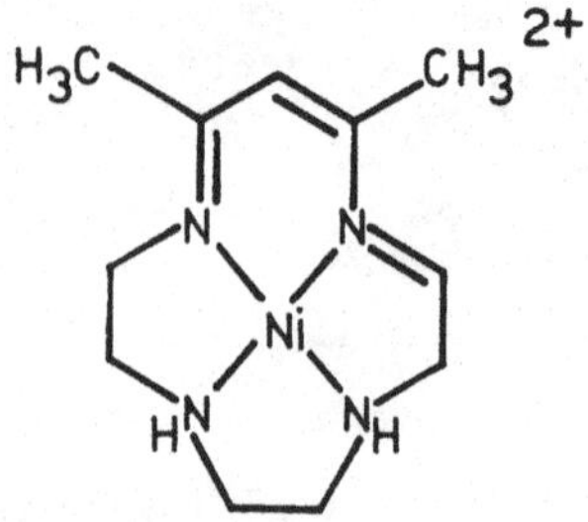

(298)

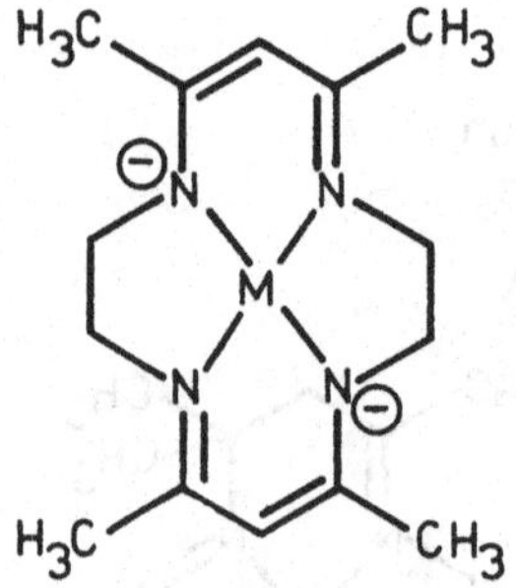

(299)

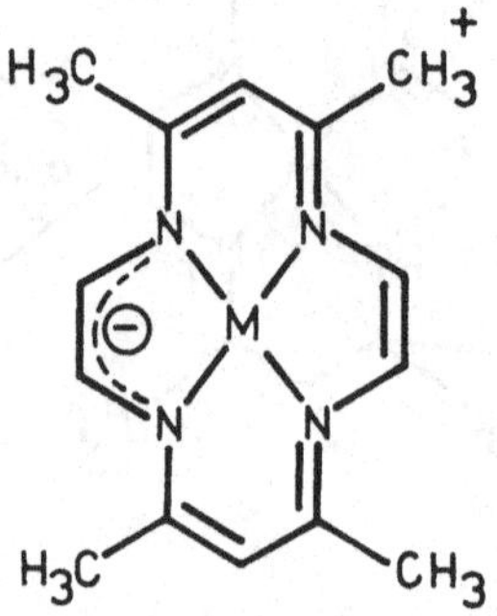

(300)

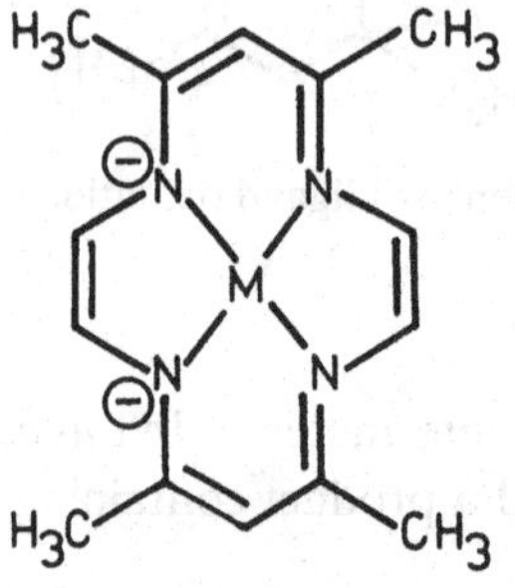

(301)

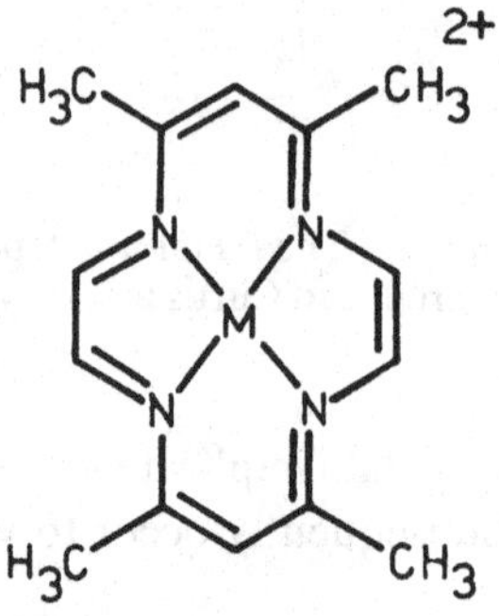

(302)

nickel complex (297) of the 13-membered macrocycle incorporating a charged, delocalized, six-membered chelate ring reacts with bromine in acetonitrile to yield the triimine product (298) (Hipp, Lindoy & Busch, 1972).

Oxidations in which unsaturation is introduced between carbon atoms in five-membered chelate rings have also been reported (Truex & Holm, 1972). Such reactions involve nickel(II) and copper(II) complexes of type (299). Treatment of these complexes with three equivalents of trityl tetrafluroborate in acetonitrile results in oxidative dehydrogenation to give 15π-electron species of type (300). Reduction of (300) with sodium borohydride in ethanol yields the (uncharged) 16π-electron species (301). Voltammetric studies indicate that (301), (300) and (302) are interconvertible by means of reversible one-electron steps with the terminal member (302) containing a stable $(4n + 2)\pi$-electron ring system.

8.4   Concluding remarks

In this chapter, the discussion has centred on the redox behaviour of the cyclic systems of a limited range of metal ions. Nevertheless, the examples are of sufficient breadth to illustrate that the redox behaviour of a particular system usually depends upon a fine balance between kinetic, thermodynamic and structural factors in which both the nature of the central metal ion and of the cyclic ligand are major influences. Of course, such considerations are not restricted to macrocyclic systems – however, the latter have provided convenient models for the elucidation of a variety of redox behaviour – much of which is of relevance to other areas and, for example, to many of the natural redox systems.

9

The natural macrocycles

In this chapter, representatives of the natural macrocycles are presented and selected model studies are also described. Emphasis will be given in the following discussion to a few of the better understood areas involving natural macrocyclic systems. Initially, the nature and function of the cyclic antibiotic category of macrocycles are discussed. Subsequently, aspects of the roles of the natural N_4-donor systems are presented.

9.1   Cyclic antibiotics and related systems

There are two general classes of naturally-occurring antibiotics which influence the transport of alkali metal cations through natural and artificial membranes. The first category contains neutral macrocyclic species which usually bind potassium selectively over sodium. The second (non-cyclic) group contains monobasic acid functions which help render the alkaline metal complexes insoluble in water but soluble in non-polar solvents (Lauger, 1972; Painter & Pressman, 1982). The present discussion will be restricted to (cyclic) examples from the first class.

Compounds such as valinomycin (303), enniatin B (304) and monactin

(303)

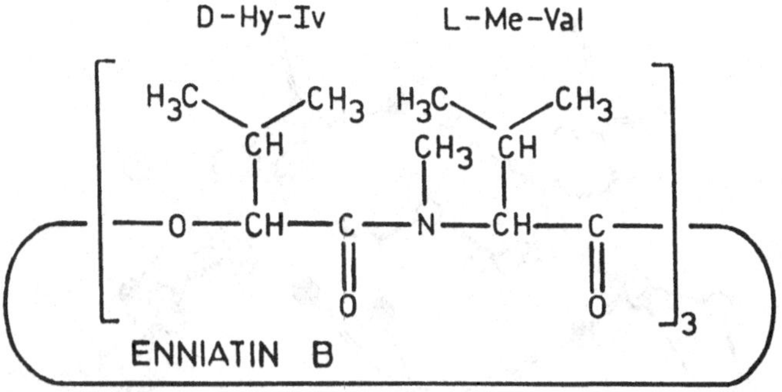

(304)

(Figure 9.1), are produced by micro-organisms and exhibit antibiotic activity. The antibiotic nature of these compounds arises because they promote changes in cellular content and membrane function thus interfering with the normal working of the cell. Valinomycin and enniatin B are depsipeptides; that is, they are cyclic compounds composed of alternating α-amino acids and α-hydroxy acids. On the other hand, monactin and the other macrotetrolides (nonactin, dinactin, trinactin and tetranactin (Figure 9.1) are ring systems which incorporate four ether and four ester groups. A feature of these compounds is that on metal complex formation, the inside of the ring is hydrophilic while the outside is hydrophobic. Complexation is thus favoured by non-polar conditions.

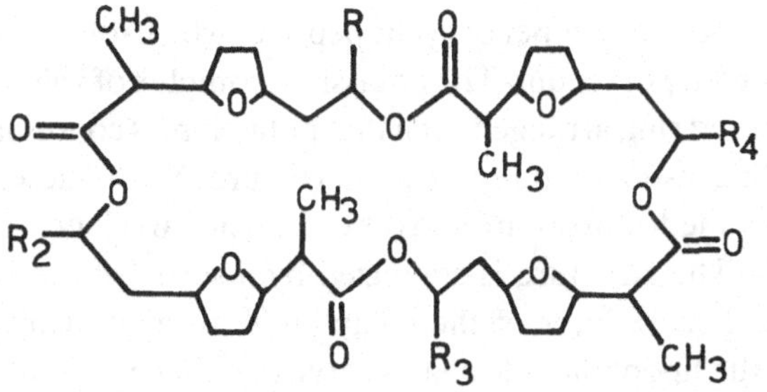

$R_1 = R_2 = R_3 = CH_3, R_4 = C_2H_5$ Monactin
$R_1 = R_2 = R_3 = R_4 = CH_3$ Nonactin
$R_1 = R_2 = CH_3, R_3 = R_4 = C_2H_5$ Dinactin
$R_1 = CH_3, R_2 = R_3 = R_4 = C_2H_5$ Trinactin
$R_1 = R_2 = R_3 = R_4 = C_2H_5$ Tetranactin

Figure 9.1. Structures of the macrotetrolide antibiotics.

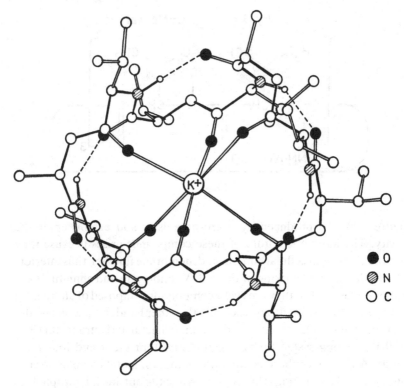

Figure 9.2. The structure of the potassium complex of valinomycin.
Hydrogens attached to carbon have been omitted.

Valinomycin is one of a number of cyclic peptides which form especially stable complexes with potassium. The potassium complex of valinomycin has the 36-membered ring arranged such that potassium is coordinated to six inward-pointing ester carbonyl groups (Figure 9.2). These donor groups form an octahedral array around the central metal (Neupert-Laves & Dobler, 1975). The structure is stabilized by a series of intra-ligand hydrogen bonds. The exterior of the complex is strongly hydrophobic and, in solution, the lipophilic side chains serve to enhance the affinity of the complex for a non-polar environment.

The macrotetrolide group (Figure 9.1) also form complexes with the alkali metals. Once again, such species preferentially complex with potassium; the potassium complex of monactin in methanol at 30°C exhibits a stability constant of 2.5×10^5 mol^{-1} dm^3 whereas the value for the sodium derivative is considerably lower at 1.1×10^3 mol^{-1} dm^3. The kinetics of complex formation by compounds of the type so far discussed with alkali metals is known to be fast. For example, members of the

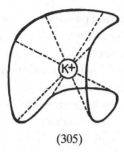

(305)

monactin series form sodium complexes in methanol whose rates of formation are very rapid indeed (c. 10^8 dm^3 mol^{-1} s^{-1}) (Diebler *et al.*, 1969). On complexation of nonactin with potassium, the cyclic molecule twists such that it resembles the seam of a tennis ball (305) (Dobler, Dunitz & Kilbourn, 1969).

The symmetrical ring enniatin (304) forms a potassium complex (Figure 9.3) whose crystal structure indicates that the cation is contained in the macrocyclic cavity and is coordinated by the six carbonyl oxygen atoms which are orientated such that three lie above and three lie below the main plane of the molecule (Dobler, Dunitz & Krajewski, 1969). Apart from those just discussed, it needs to be noted that a range of other structures of antibiotic molecules and their metal complexes have been determined (Hilgenfeld & Saenger, 1982).

Metal-ion transport through biological membranes

Biological membranes, such as those surrounding cells, are made up of lipids and proteins exhibiting a more or less ordered arrangement. These membranes are about 70 Å thick and allow ions and nutrients to

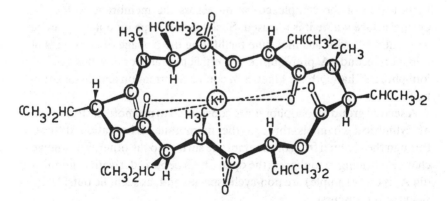

Figure 9.3. The potassium complex of enniatin B (from Fenton, 1977).

pass into the cell while permitting passage out of a range of species which are either unwanted or produced for use elsewhere. Such membranes do not permit ions (and other species) to pass through indiscriminately; rather, they are selective, resulting in the ionic composition inside the cell being quite different to that in the surrounding medium. For example, the inside of a cell contains a much larger potassium ion concentration than sodium even though the reverse situation normally applies outside the cell. The presence of the sodium/potassium gradient across the cell membrane results in the generation of a potential difference between the inside and outside of the cell. The existence of this potential has important implications for physiological processes such as the transmission of the action potential along the membranes of nerve cells and the contraction of muscles.

The compounds just discussed have all been implicated in alkali metal-ion transport and related phenomena in biological systems. Substances such as these, which are capable of carrying ions across a hydrophobic membrane, are called **ionophores**.

Ionophores are necessary since the lipid components of biological membranes tend to be orientated such that their polar groups face the membrane surfaces while the non-polar hydrocarbon portions occupy the membrane interior. The hydrophobic nature of the centre of the membrane thus acts as a barrier to the passage of ions such as sodium or potassium.

There appear to be two major ways by which ionophores aid ions to cross membrane barriers. Ionophores such as valinomycin and nonactin enclose the cation such that the outside of the complex is quite hydrophobic (and thus lipid-soluble). The transport behaviour thus involves binding of the cation at the membrane surface by the antibiotic, followed by diffusion of the complexed cation across the membrane to the opposite surface where it is released. Such 'carrier' type ionophores can be very efficient, with one molecule facilitating the passage of thousands of ions per second. A prerequisite for efficient transport by this type of ionophore is that both the kinetics of complex formation and dissociation be fast.

A second group of ionophores are considered to promote the formation of cylindrical channels through the membrane. The cation diffuses through the channel from one membrane surface to the other. The known channel-forming ionophores (the open-chain peptide derivative, gramicidin A, is one example) are non-cyclic species and, as such, lie outside the scope of this discussion.

There are two major driving forces involved in natural membrane transport (Racker, 1979). The first is called facilitated diffusion and involves carrier-assisted 'passive' transport. For this, no energy source is required. Alternatively, the cations may be transported against a concentration gradient. This process requires energy, which is ultimately obtained by hydrolysis of adenosine triphosphate (ATP) – such a process occurs in the so-called 'sodium pump' of biological systems.

The crowns as model carriers. Many studies involving crown ethers and related ligands have been performed which mimic the ion-transport behaviour of the natural antibiotic carriers (Lamb, Izatt & Christensen, 1981). This is not surprising, since clearly the alkali metal chemistry of the cyclic antibiotic molecules parallels in many respects that of the crown ethers towards these metals. As discussed in Chapter 4, complexation of an ion such as sodium or potassium with a crown polyether results in an increase in its lipophilicity (and a concomitant increase in its solubility in non-polar organic solvents). However, even though a ring such as 18-crown-6 binds potassium selectively, this crown is expected to be a less effective ionophore for potassium than the natural systems since the two sides of the crown complex are not as well-protected from the hydrophobic environment existing in the membrane.

Typically, transport experiments have been performed using a U-tube apparatus in which a solvent such as chloroform, containing the macrocyclic carrier, is placed in the tube so that it separates two aqueous phases: the source phase containing the metal ion(s) to be transported and the receiving phase into which the transported ions are deposited. A diagrammatic representation of a liquid membrane system is shown in Figure 9.4.

It should be noted that selective 'carrier facilitated' transport experiments involving the crown ethers and their derivatives have not been

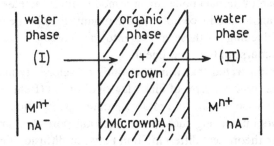

Figure 9.4. Schematic representation of carrier-mediated metal-ion transport through a liquid membrane (A = anion).

restricted to the alkali metals – a range of other metals including the alkaline earths, Zn(II), Cd(II), Ag(I) and Pb(II) have also been studied in some detail (Izatt *et al.*, 1983; Izatt, Hawkins, Christensen & Izatt, 1985).

When neutral macrocyclic ligands such as 18-crown-6 bind to the metal cation, a positively-charged species results. Hence on crossing a hydrophobic liquid membrane, a counter-ion must accompany the complexed cation if electrical neutrality is to be maintained. The nature of the anion has a large effect on the cation fluxes in systems such as these. For example, the rate of transport of potassium across a chloroform barrier using dibenzo-18-crown-6 as carrier shows an increase of eight orders of magnitude on substituting the large picrate ion for fluoride ion in the source phase (Lamb *et al.*, 1980).

Other factors influencing the rate of metal-ion transport across artificial membranes have been identified. As might be expected, such transport is dependent on the interplay of several factors. For example, as briefly mentioned already in Chapter 4, it is clear that the strength of complexation of the cation by the carrier must be neither too high nor too low if efficient transport is to be achieved. If the stability is low, then uptake of the metal ion from the source phase will be inhibited. Conversely, for those cases where highly stable complexes are formed, there will be a reluctance by the carrier to release the cation into the receiving phase.

In many of the experiments performed so far, the driving force for transport has been the metal-ion activity gradient on passing from the source phase to the receiving phase. Thus, in such cases, transport through liquid membranes takes place only when the concentration of metal ion is high enough in the source phase to result in complexation of the carrier and proceeds as long as metal-ion loss from the carrier occurs at the receiving interface.

In a number of studies, it has been found advantageous to increase the lipophilicity of the macrocyclic carrier by appending long-chain alkyl groups to the macrocycle backbone. Such a modification increases the solubility of the carrier and its complex in the organic membrane phase while inhibiting 'bleeding' of these species into the source or receiving phases during the course of the experiment.

It has been demonstrated that transport rate and selectivity may be modelled using the basic concepts of Fick's law of diffusion (Behr, Kirch & Lehn, 1985). Analyses of this type allow a greater appreciation of the interplay of factors influencing such membrane transport phenomena, and enable a clear theoretical differentiation between diffusion-limited and complexation rate-limited cases.

A variety of other ligands, such as the calixarenes (see Chapter 5), have also been used as selective carriers across liquid membranes (Izatt *et al.*, 1983). These latter ligands are very suitable for alkali metal transport because of their low water-solubility and their ability to form neutral complexes through proton loss. This latter ability mimics the similar behaviour of the open-chain (charged) category of natural antibiotics (Taylor, Kauffman & Pfeiffer, 1982) mentioned earlier in this chapter.

9.2   Natural corrin, porphyrin and related systems

As a class, metal-ion derivatives of tetrapyrrole macrocyclic rings, such as the corrins or porphyrins (see Chapter 1 for the parent ring structures), are of major biological importance.

Metal complexes of the porphyrins have been studied for many years. Such attention is not surprising, since particular derivatives play a central role in photosynthesis, dioxygen transport and storage as well as other fundamental processes such as electron transfer (Smith, 1975; Dolphin, 1978–9). Indeed, there are few compounds found in nature which can compare with the diversity of biochemical functions exhibited by the porphyrins.

On coordination, the porphyrin macrocycle loses two protons (to yield a neutral complex when the central metal ion is divalent). The extensive electron delocalization throughout the ligand will normally extend to the central metal when the latter is covalently bound to the porphyrin. As expected, such complexes are extremely stable; this is undoubtedly important to the biological role of these complexes.

In comparison to the porphyrins, the corrin nucleus contains one less atom in its innermost ring (that is, it contains a 15-membered ring) and, on coordination, only one NH proton is lost to give the macrocycle a single negative charge. A cobalt corrin complex occurs as part of the structure of vitamin B_{12}.

Vitamin B_{12}

The structure of vitamin B_{12} coenzyme, otherwise known as 5'-deoxyadenosine cobalamin, is given by (306). This species acts as a cofactor for methylmalonyl coenzyme A mutase in the conversion of methylmalonyl coenzyme A into succinyl coenzyme A. This vitamin B_{12} species is a novel, naturally-occurring organometallic compound incorporating a cobalt-carbon linkage in its structure. As well as (306), the corresponding derivative with $R = CH_3$ also acts as a cofactor in enzymic processes. Formally, these complexes can be considered to contain Co(III) and a bound carbanion although it is noted that lower oxidation species are undoubtedly important in their biological function.

(306)

Vitamin B_{12} itself is cyanocobalamin in which the sixth group is a cyanide ion. The presence of cyanide arises from the isolation procedure. Other derivatives are B_{12a} (aquocobalamin) and B_{12c} (nitrocobalamin). The B_{12a} form may be reduced first to B_{12r} and then to B_{12s}. B_{12r} contains low-spin Co(II), while B_{12s} appears to contain Co(I).

Many model systems which mimic both the redox behaviour [for example, ready reduction to Co(I) species] and the alkyl binding ability of vitamin B_{12} derivatives have been investigated. The most studied of these has involved bis(dimethylgloximato)cobalt systems of type (307), known as the 'cobaloximes' (Bresciani-Pahor *et al.*, 1985). Other closely related

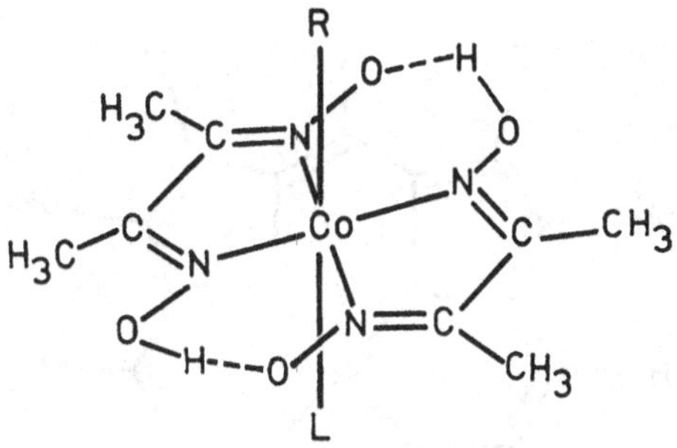

(307)

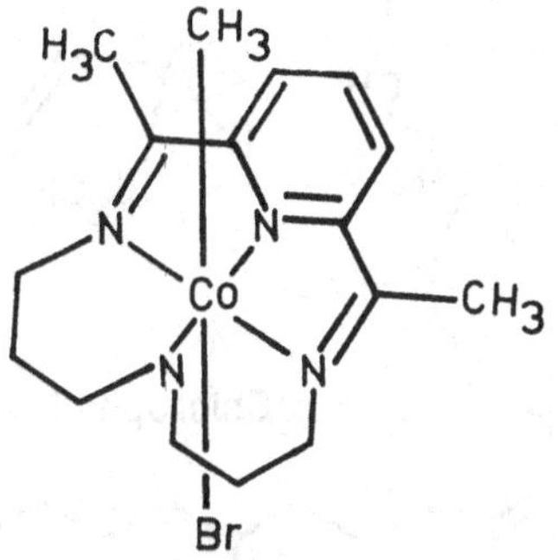

(308)

derivatives have also been investigated (Parker *et al.*, 1986). Indeed, a variety of other Co(III) models have been investigated, including macrocyclic complexes (Farmery & Busch, 1970; Ochai & Busch, 1968) such as (308). The cobalt-alkyl bond in such species is quite stable and much interest has centred on the factors influencing this stability. It has been demonstrated that the degree of unsaturation in the equatorial ligand(s) is important in this respect.

Overall, the study of B_{12} models has brought about a very much fuller understanding of the chemistry of the natural systems.

Chlorophyll

The chlorophyll molecule (309) is involved in initiating photosynthesis in green plants and contains magnesium coordinated to a partially reduced porphyrin (namely, a chlorin derivative). Life relies ultimately on the unique redox and electron transfer abilities of the chlorophylls which are necessary for the conversion of light to chemical energy. Chlorophyll mainly absorbs light from the far red region of the spectrum

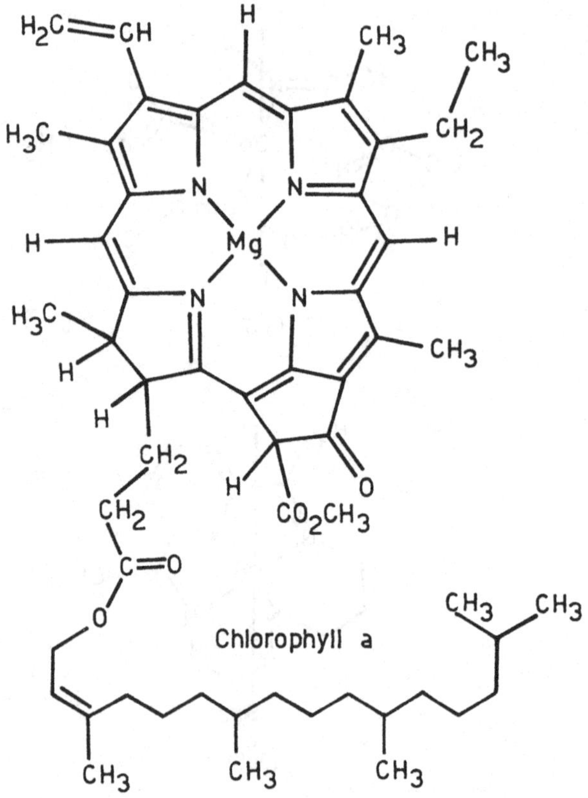

Chlorophyll a

(309)

(green light is transmitted) with the frequency of absorption depending upon the substituents present on the chlorin ring.

Reduced porphyrin derivatives are not only involved in photosynthesis; for example, a derivative of this type (isobacteriochlorin) is also concerned with the binding of soil nitrite and its reduction to ammonia.

Photosynthesis in green plants involves the oxidation of water to dioxygen, followed by reduction of carbon dioxide to yield glucose. That is, the latter process involves the synthesis of glucose from CO_2 and H_2O via a photo-initiated reaction.

Although studied in great detail, the action of chlorophyll is still not fully understood; however, steady progress towards a more complete understanding has taken place over the past several decades. Two photosynthetic systems (photosystems I and II) are present in green plants – each incorporating a different chlorophyll type. When a photon is absorbed by a chlorophyll molecule, its energy is transformed and becomes available for a complex series of redox processes which are initiated. A number of individual biological redox systems including the cytochromes (see later) and ferredoxins have been shown to be involved in the subsequent electron transfer reactions (Livorness & Smith, 1982). These reactions lead to the production of O_2 as well as to the generation of $NADPH_2$. The latter plays an important role in the conversion of CO_2 to carbohydrate. The formation of carbohydrate provides a means of energy storage: although sunlight provides the primary energy source, a storage mechanism remains necessary if the system is to be maintained in the dark.

Haem proteins and O_2 transport and storage

Haem proteins are a class of compounds of fundamental impor-tance to both respiratory and other biological processes (Sykes, 1982). Fe(II) protoporphyrin IX (310) (the 'prosthetic' group) and associated protein chains ('globin' chains) occur in most vertebrates. The latter chains show inter and intra species variation, although sometimes differ-ing by as little as one or two amino acid residues. In blood, the haem protein is haemoglobin (Hb) and, in muscle, it is myoglobin (Mb). Thus, haemoglobin is concerned with the *in vivo* absorption of dioxygen from air or water and its transport via the blood to muscle tissue. On the other hand, myoglobin is involved in dioxygen storage until energy release is required. Hence, in the complete sequence, the deoxygenated form of haemoglobin binds molecular oxygen and transports it from the lungs to muscle tissue where it is stored by myoglobin until required for metabolic action (upon which, after being transported to the mitochondria, energy

(310)

Table 9.1. Comparison of the properties of haemoglobin and myoglobin.

	Haemoglobin	Myoglobin
Function	Transport	Storage
Molecular weight	64 450	17 500
Number of subunits	$4(2\alpha, 2\beta)$	1
Iron content	4	1
Deoxy form	4 Fe(II)	1 Fe(II)
Fe : O_2 ratio	1 : 1	1 : 1

is released by reaction with glucose to produce, ultimately, CO_2 and water).

Myoglobin has also been implicated in the rate of diffusion of dioxygen through the cell matrix to the mitochondria. Haemoglobin also has a second role: to bind the CO_2 produced and transport it back to the lungs. Binding in this case appears to involve amine groups at the beginning of the polypeptide chains with the formation of carbamino ($-NHCO_2^-$ groups.

Table 9.1 gives a comparison of the respective properties of haemoglobin and myoglobin.

In myoglobin, the Fe(II) porphyrin is held within a cleft of the associated protein molecule (150–160 amino acid residues long) by means of a considerable number of hydrophobic interactions together with a co-valent link between the Fe(II) and an imidazole group from a histidine residue in the protein. In the absence of dioxygen, myoglobin contains a five-coordinate Fe(II) in its high-spin state. The iron is positioned above the N_4-donor plane of the porphyrin towards the axial imidazole group (Takano, 1977). On O_2 uptake, the dioxygen binds to the sixth coordination site of the Fe(II) and there is a spin-state change to low-spin; the Fe(II) now lies closer to the N_4-donor plane (Phillips, 1980). The O_2 molecule coordinates in a bent, end-on geometry and is hydrogen-bonded to a nearby imidazole group (Phillips & Schoenborn, 1981).

Related behaviour to that just described occurs in haemoglobin, although the latter is a tetramer consisting of four subunits each containing an iron porphyrin. The latter are located within hydrophobic pockets in the globin portion of the molecule (which, in this case, is composed of four linked chains – consisting of two α and two β chains, which differ in their respective amino acid compositions).

Once again, each haem is also bound to an imidazole of a histidine residue (called the 'proximal' imidazole) from the surrounding globin (Figure 9.5). The iron is displaced approximately 0.6 Å from the Fe–N_4

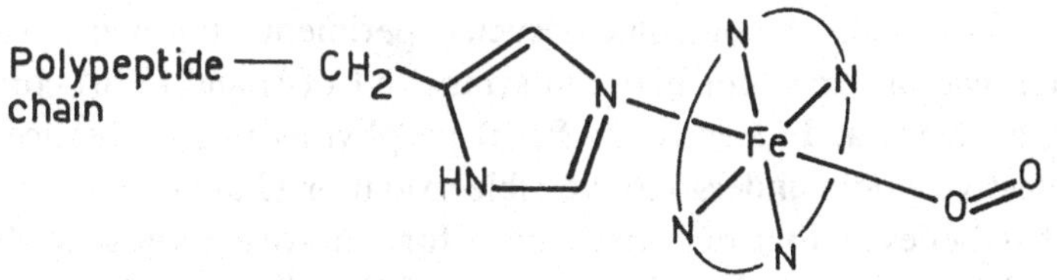

Figure 9.5. Attachment of the haem group to the globin chain via a coordinated histidine group.

plane towards the imidazole group in the deoxy form (Fermi, 1975; Ten Eyck & Arnone, 1976); as mentioned in earlier chapters, such displacement is typical of a square-pyramidal complex. Within the hydrophobic pocket, O_2 binds to the Fe(II) on the 'distal' side of the porphyrin ring causing the iron to move back into the coordination plane with an accompanying change of spin-state (to low-spin). Since the low-spin form of Fe(II) has a smaller radius, this may result in a better fit of the iron in the macrocyclic cavity. On reaction with dioxygen, haemoglobin is rapidly converted from its T (tense) state to its R (relaxed) state.

Haemoglobin is more than a passive dioxygen carrier – the dioxygen binding curve is not hyperbolic (unlike that for myoglobin) but rather is sigmoidal reflecting 'cooperative' binding between the four haem sites. Thus the dioxygen affinity of haemoglobin increases with increasing oxygen concentration. Myoglobin has a stronger affinity for O_2 than does haemoglobin at low oxygen partial pressures and in the presence of high carbon dioxide concentrations. This aids dioxygen transfer from haemoglobin to myoglobin when such conditions prevail.

Even though the iron atoms are separated in haemoglobin by about 25 Å, communication between them is still able to occur and this has been postulated to involve a 'trigger' mechanism (Perutz, 1971). The trigger is the movement of the proximal histidine as dioxygen binds to (or is released from) the Fe(II) and results in interconversion between the T and R structures. This movement causes a conformational change which is transmitted through the protein to the other iron sites. X-ray studies indicate that relative shifts of up to 6 Å at subunit interfaces occur between the T and R states (Perutz, 1978).

Models for haemoglobin and myoglobin. It has been known for a long time that a range of synthetic Co(II) complexes are able to bind reversibly dioxygen in a manner related to the natural systems (Basolo, Hoffman & Ibers, 1975; McLendon & Martell, 1976). However, similar complexes of iron do not normally function as O_2 carriers – rather oxidation to Fe(III) species is usually the outcome of such experiments. For many years, the difference in behaviour between synthetic Co(II) and Fe(II) complexes was an enigma. Indeed, even Fe(II) porphyrins in the absence of the protective globin undergo irreversible oxidation (James, 1978).

It is believed that one mechanism for auto-oxidation of such Fe(II) complexes involves the initial binding of O_2 followed by a rapid redox process involving dimerization via an oxygen-bridged species to yield a Fe(III) μ-oxo dimer as the final product:

$$4Fe(II) + O_2 \rightarrow 2Fe(III)-O-Fe(III).$$

In model studies involving Fe(II) species, three broad approaches have been used to mitigate the problem of autoxidation of the iron (Hay, 1984). These are: (i) the use of low temperatures so that the rate of oxidation becomes very slow; (ii) the synthesis of ligands containing steric barriers such that dimerization of the iron complex is inhibited, and (iii) immobilization of the iron complex on a solid surface such that dimerization once again will not be possible.

In one study of the first type, the iron complex (311) was investigated as a model for the myoglobin active site (Chang & Traylor, 1973). This complex was demonstrated to bind dioxygen reversibly in dichloromethane at $-45\,°C$; however, its efficiency decreases due to irreversible oxidation as the number of cycles increases. Like myoglobin and haemoglobin, this species binds carbon monoxide more strongly than dioxygen (in the natural systems, the corresponding strong affinity for CO is the reason for the latter's toxicity). Other studies at low temperatures have demonstrated that the covalent attachment of an imidazole derivative is not essential for reversible oxygen binding under such conditions (Wagner & Kassner, 1974; Brinigar & Chang, 1974; Anderson, Weschler & Basolo, 1974; Almog et al., 1974). Nevertheless, the presence of this base still appears to aid O_2-binding (Brinigar & Chang, 1974).

A number of model studies of the second type have also been investigated. In these, the iron complex provides a hydrophobic cavity so positioned that it surrounds the bound O_2 molecule. These include the 'lacunar' complexes mentioned in Chapter 3 (Goldsby, Beato & Busch, 1986); particular complexes of which are quite efficient reversible carriers. A further (early) example of this type was (312) which encloses the dioxygen in a rudimentary cavity. However, this species only binds dioxygen reversibly at low temperatures (Baldwin & Huff, 1973).

Other studies have been based on synthetic systems incorporating haem moieties (Baldwin & Perlmutter, 1984). One elegant study of this type involves the Fe(II) complex of the 'picket fence' ligand (313)

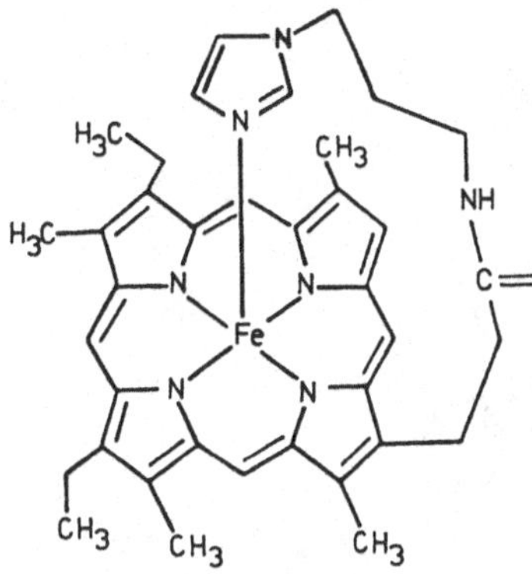

(311)

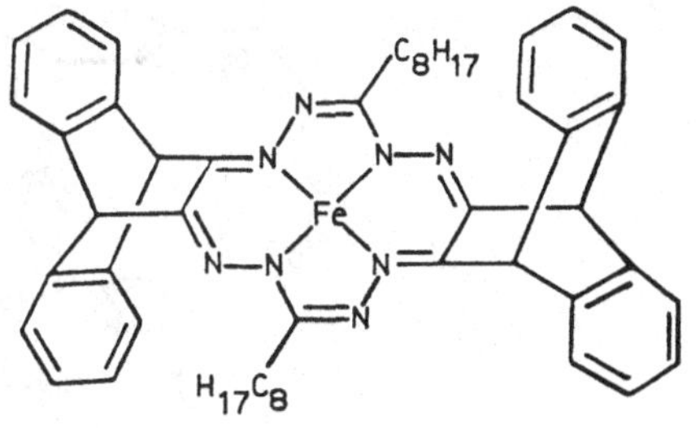

(312)

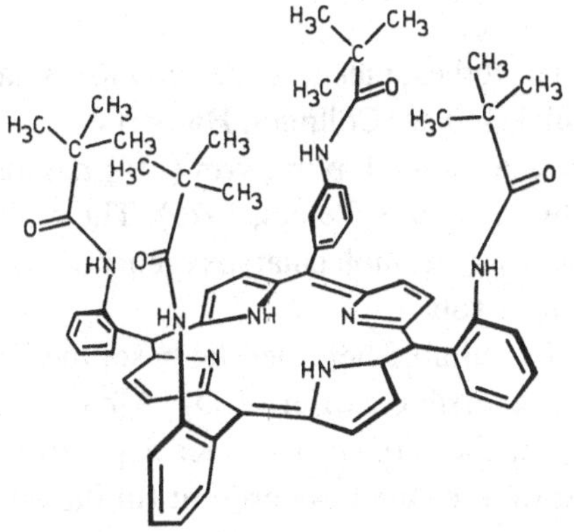

(313)

(Collman *et al.*, 1974). This complex undergoes reversible oxygenation *at room temperature* in the presence of a base such as *N*-methylimidazole and, under defined conditions, extremely long half-lives for the oxy form have been reported. The dioxygen adduct containing *N*-methylimidazole has been isolated and characterized by X-ray analysis (Collman *et al.*, 1974; Collman, 1977). Clearly, the presence of the bulky substituents is sufficient to inhibit oxygen-bridged dimerization in this system. This model shares many similarities with the natural systems. In the oxy form, the diamagnetic Fe(II) occupies the plane of the porphyrin ring, while the iron-oxygen bond appears to involve substantial double bond character (Collman, *et al.*, 1974; Collman, Cagne, Gray & Hare, 1974). The latter is bent (with an Fe–O–O angle of 136°C); such bent 'end-on' coordination had previously been predicted to occur in oxy-haemoglobin (Pauling, 1964) and was subsequently confirmed by X-ray studies (Shaanan, 1982).

(314)

Following these initial studies, picket fence derivatives have been used in a range of other model studies (Collman, Halbert & Suslick, 1980).

In a further study the 'capped' porphyrin (314) has also been synthesized (Almog, Baldwin, Dyer & Peters, 1975). The cavity in this case is large enough to incorporate molecular oxygen but will exclude other small molecules (such as solvent).

In the presence of a suitable heterocyclic base, the Fe(II) complex of this system is also a reversible carrier of O_2 (Almog, Baldwin & Huff, 1975). The stability of the dioxygen adduct depends largely upon the nature and concentration of the base present (in the absence of a base, rapid dimerization and autoxidation occurs).

In a typical experiment, the Fe(II) derivative of (314) rapidly binds dioxygen in pyridine; subsequent deoxygenation may be effected by freeze–thawing. Several such cycles can be performed with little deterioration of the system. Moreover, the O_2 adduct in pyridine has a half-life of about 20 hours. Following this initial success, the O_2-binding properties of a number of other related capped porphyrin derivatives have been investigated (Baldwin & Perlmutter, 1984).

The third group of studies involves attachment of the iron complexes to solid substrates in order to inhibit formation of bridged species. In a very early study, dioxygen was found to bind reversibly to haem diethyl ester embedded in a mixture of polystyrene and 1-(2-phenylethyl)imidazole (Wang, 1958).

Attachment of the Fe(II) complex of tetraphenylporphyrin to a modified silica gel support also resulted in an efficient dioxygen carrier (Basolo, Hoffman & Ibers, 1975). In this system the silica gel was modified by direct attachment of imidazoyl-propyl groups to its surface.

The Fe(II) complex of tetraphenylporphyrin was then attached to the modified substrate by coordination of the immobilized base in an axial site. The resulting five-coordinate complex binds dioxygen reversibly.

The cytochromes

The cytochromes are another group of haem proteins found in all aerobic forms of life. Cytochromes are electron carriers involving a Fe(II)/Fe(III) redox system. They are a crucial part of the electron transfer reactions in mitochondria, in aspects of the nitrogen cycle, and in enzymic processes associated with photosynthesis.

The large number of cytochromes identified contain a variety of porphyrin ring systems. The classification of the cytochromes is complicated because they differ from one organism to the next; the redox potential of a given cytochrome is tailored to the specific needs of the electron transfer sequences of the particular system. The cytochromes are one-electron carriers and the electron flow passes from one cytochrome type to another. The terminal member of the chain, cytochrome c oxidase, has the property of reacting directly with oxygen such that, on electron capture, water is formed:

$$O_2 + 4H^+ + 4e^- \rightarrow 2H_2O.$$

Cytochrome c. A range of cytochrome c molecules exist and collectively they are the most studied group of cytochromes. They contain the haem group covalently linked to the protein via a thioether group; in most cases they are low-spin and incorporate either histidine and methionine or two histidine groups as the axial ligands for the iron. Cytochrome c can be isolated from the mitochondrial membrane; the structures of the Fe(II) and Fe(III) forms of horse heart cytochrome c have been determined (Dickerson & Timkovich, 1975). The single peptide chain forms an approximately spherical mass around the haem group with only about 20% of the edge of the porphyrin ring being exposed.

Cytochrome c is responsible for accepting an electron from cytochrome c_1 and transferring it to cytochrome c oxidase. The electron transfer reaction may occur via the exposed portion of the ring or by 'tunnelling' through the protein (and involving an outer-sphere mechanism). The details of this process have not been fully elucidated and have remained the focus of much research.

Cytochrome c oxidase. Cytochrome c oxidase is an enzyme that occurs in the inner mitochondrial membrane and, as we have seen, catalyzes the four-electron reduction of dioxygen to water as the final reaction in the mitochondrial electron transport chain. Thus, this enzyme accepts four electrons from cytochrome c and transfers them to bound dioxygen to yield water; the process is of physiological importance since it is accompanied by a release of energy which may be stored by conversion of ADP to ATP. The enzyme contains seven peptide subunits, two haems and two copper ions. It contains two different cytochromes (a and a_3).

Although several proposals have been put forward, the detailed mode of action of this enzyme remains uncertain. However, there is evidence that the O_2 molecule binds to one of the cytochromes while the other serves as an electron shuttle to the dioxygen complex (Malmstrom, 1980). An interesting feature is that strong antiferromagnetic coupling appears to occur between Fe(III) and Cu(II) in this system (Van Gelder & Beinert, 1969); this has provided motivation for the synthesis of model compounds in an attempt to observe related coupling. Although such mixed-metal systems have been synthesised, their behaviour does not always follow closely that of the natural system (Gunter *et al.*, 1980). However, a dinuclear system of type FeCu(O)(H$_2$O)(OAc)L (L = 315) incorporating high-spin Fe(III) (in the porphyrin ring) joined to Cu(II) via a Fe–O–Cu bridge, does exhibit significant coupling between the metal centres. In this respect, this system mimics in part the postulated situation in the enzyme (Chang, Koo & Ward, 1982).

(315)

Cytochrome P–450. Cytochrome P-450 enzymes consist of a large number of haem-containing mono-oxygenases which catalyze aliphatic and aromatic hydroxylations, epoxidations, as well as other oxidation reactions; thus, these enzymes are able to cleave aromatic C–H bonds and also unactivated alkane C–H bonds both regio- and stereo-specifically. Their name derives from the fact that the carbon monoxide adducts show absorption bands at 450 nm. It should be noted that the 'cytochrome' in the name may mislead, as the enzymic functions are very different from those of the cytochromes associated with the electron transport chain. Indeed, the P–450 enzymes appear to have a number of functions – including the hydroxylation of lipid-soluble drugs, pesticides and other foreign substances (xenobiotics) which find their way into animals, for example, through diet, (Hay, 1984); hydroxylation results in increased water solubility and a consequent greater ease of excretion through the kidneys. Thus, these enzymes are an important part of the body's detoxification system.

Since first isolated in the 1950s much attention has been given to the mechanistic and structural aspects of this class of enzymes. Cytochrome P-450 is composed of a single haem group and a single protein chain with the iron appearing to have one axial site occupied by a cysteine sulfur atom; the resting state of the enzyme contains low-spin Fe(III). The nature of the ligand occupying the sixth site is not known with certainty. In any case, this ligand is lost on forming a reactive five-coordinate, high-spin Fe(III) species in the presence of the substrate. The Fe(III) species is reduced to a high-spin Fe(II) state and is then converted to the corresponding low-spin complex containing bound O_2. During hydroxylation, activated dioxygen reacts with the substrate (and is accompanied by the reductive cleavage of the O–O bond). Thus, cytochrome P-450 appears to be involved in a catalytic cycle of the type outlined in Figure 9.6.

Once again, models for a number of the compounds illustrated in the cycle have been synthesised (Chottard, Schappacher, Ricard & Weiss, 1984). In particular, major interest has been shown in elucidating how the enzyme can activate a dioxygen molecule while oxygen carriers such as myoglobin or haemoglobin, with related structures, form stable dioxygen complexes. With respect to this, axial ligation by cysteine appears to be an important difference in the P-450 system relative to the oxygen-carrying systems (Groves, Watanabe & McMurray, 1983; Smegal & Hill, 1983).

In a number of model studies it has been demonstrated that synthetic porphyrin complexes are able to be oxidized to oxo-metalloporphyrin complexes which will, in turn, oxidize organic substrates of the type mentioned previously. Thus, in an early investigation it was shown that a synthetic Fe(III) porphyrin will catalyze oxygenation of hydrocarbons using iodosylbenzene as the reduced-oxygen source (Groves, Nemo & Myers, 1979). Following this report, a number of related cytochrome

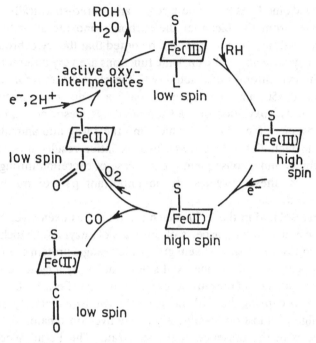

Figure 9.6. Proposed cytochrome P-450 catalytic cycle (S = cysteine).

P-450 models have been investigated (Groves, Kruper & Haushalter, 1980; Groves, Watanabe & McMurray, 1983; Smegal & Hill, 1983; Groves & Gilbert, 1986). These systems consist of a Fe(III) or Mn(III) porphyrin together with an oxygen source which may be a single oxygen donor such as iodosylbenzene or hypochlorite, or may be a combination of molecular oxygen and a reducing agent. As in the 'activated' form of the P-450 enzyme, in all cases it appears that the active species generated is a high-valent, oxo-porphyrin complex.

References

Chapter 1

Abid, K. K., Fenton, D. E., Casellato, U. & Vigato, P. A. (1984). *Journal of The Chemical Society, Dalton Transactions*, 351–4.

Alcock, N. W., Herron, N. & Moore, P. (1978). *Journal of The Chemical Society, Dalton Transactions*, 1282–8.

Barefield, E. K. & Wagner, F. (1973). *Inorganic Chemistry*, **12**, 2435–9.

Bishop, M. M., Lewis, J., O'Donoghue, T. D., Raithby, P. R. & Ramsden, J. N. (1980). *Journal of The Chemical Society, Dalton Transactions*, 1390–6.

Black, D. St. C. & McLean, I. A. (1968). *Journal of The Chemical Society, Chemical Communications*, 1004.

Bosnich, B., Poon, C. K. & Tobe, M. L. (1965). *Inorganic Chemistry*, **4**, 1102–8.

Busch, D. H. (1978). *Accounts of Chemical Research*, **11**, 392–400.

Constable, E. C., Khan, F. K., Lewis, J., Liptrot, M. C. & Raithby, P. R. (1985). *Journal of The Chemical Society, Dalton Transactions*, 333–5.

D'Aniello, M. J., Mocella, M. T., Wagner, F., Barefield, E. K. & Paul, C. (1975). *Journal of the American Chemical Society*, **97**, 192–4.

Drew, M. G. B., Cabral, J. DeO., Cabral, M. F., Esho, F. S. & Nelson, S. M. (1979). *Journal of The Chemical Society, Chemical Communications*, 1033–5.

Goedken, V. L., Pluth, J. J., Peng, S. M. & Bursten, B. (1976). *Journal of the American Chemical Society*, **98**, 8014–21.

Henrick, K., Tasker, P. A. & Lindoy, L. F. (1985). *Progress in Inorganic Chemistry*, **33**, 1–58.

Herron, N. & Moore, P. (1979). *Inorganica Chimica Acta*, **36**, 89–96.

Hoard, J. L. (1975). In *Porphyrins & Metalloporphyrins*, chapter 8, ed. K. M. Smith. Elsevier, Amsterdam.

Horner, L., Walach, P. & Kunz, H. (1978). *Phosphorus and Sulfur*, **5**, 171–84.

Imajo, S., Nakanishi, K., Roberts, M. & Lippard, S. J. (1983). *Journal of the American Chemical Society*, **105**, 2071–3.

Kauffmann, T. & Ennen, J. (1981). *Tetrahedron Letters*, 5035–8.

Klaehn, D.-D., Paulus, H., Grewe, R. & Elias, H. (1984). *Inorganic Chemistry*, **23**, 483–90.

Koyama, H. & Yoshino, T. (1972). *Bulletin of the Chemical Society, Japan*, **45**, 481–4.

Lai, T. F. & Poon, C. K. (1976). *Inorganic Chemistry*, **15**, 1562–6.

Lindoy, L. F. & Busch, D. H. (1969). *Journal of the American Chemical Society*, **91**, 4690–3.

Mealli, C., Sabat, M., Zanobini, F., Ciampolini, M. & Nardi, N. (1985). *Journal of The Chemical Society, Dalton Transactions*, 479–85.

Micheloni, M., Paoletti, P., Burki, S. & Kaden, T. A. (1982). *Helvetica Chimica Acta.*, **65**, 587–94 and references therein.

Micheloni, M., Paoletti, P., Siegfried-Hertli, L. & Kaden, T. A. (1985). *Journal of The Chemical Society, Dalton Transactions*, 1169–72.

Newkome, G. R. & Lee, H. W. (1983). *Journal of the American Chemical Society*, **105**, 5956–7.

Ogawa, S. & Shiraishi, S. (1980). *Journal of The Chemical Society, Perkin Transactions* I, 2527–30.

Pett, V. B., Diaddario, L. L., Dockal, E. R., Corfield, P. W., Ceccarelli, C., Glick, M. D., Ochrymowycz, L. A. & Rorabacher, D. B. (1983). *Inorganic Chemistry*, **22**, 3661–70.

Richman, J. E., Atkins, T. J. (1974). *Journal of the American Chemical Society*, **96**, 2268–70.

Robertson, G. & Whimp, P. (1988). *Personal communication*.

Truex, T. J. & Holm, R. H. (1972). *Journal of the American Chemical Society*, **94**, 4529–38.

Vallee, B. L. & Williams, R. J. P. (1968). *Proceedings of the National Academy of Science (USA)*, **59**, 498–505.

Wagner, F. & Barefield, E. K. (1976). *Inorganic Chemistry*, **15**, 408–17.

Zuckman, S. A., Freeman, G. M., Troutner, D. E., Volkert, W. A., Holmes, R. A., Van Derveer, D. G. & Barefield, E. K. (1981). *Inorganic Chemistry*, **20**, 2386–9.

Chapter 2

Ansell, W. G., Cooper, M. K., Dancey, K. P., Duckworth, P. A., Henrick, K., McPartlin, M. & Tasker, P. A. (1985). *Journal of The Chemical Society, Chemical Communications*, 439–41.

Black, D. St. C., Bos, C. H., Vanderzalm, C. H. & Wong, L. C. H. (1979). *Australian Journal of Chemistry*, **32**, 2303–11.

Blinn, E. L. & Busch, D. H. (1968). *Inorganic Chemistry*, **7**, 820–4.

Busch, D. H., Farmery, K., Goedken, V., Katovic, V., Melnyk, A. C., Sperati, C. R. & Tokel, N. (1971). *Advances in Chemistry Series*, **100**, 44–78.

Comba, P., Curtis, N. F., Lawrance, G. A., Sargeson, A. M., Skelton, B. W. and White, A. H. (1986). *Inorganic Chemistry*, **25**, 4260–7.

Constable, E. C., Khan, F. K., Lewis, J., Liptrot, M. C. & Raithby, P. R. (1985). *Journal of The Chemical Society, Dalton Transactions*, 333–5.

Cummings, S. C. & Busch, D. H. (1970). *Journal of the American Chemical Society*, **92**, 1924–9.

Cummings, S. C. & Sievers, R. E. (1970). *Inorganic Chemistry*, **9**, 1131–6.

Curtis, N. F. (1960). *Journal of The Chemical Society*, 4409–17.

Curtis, N. F., Curtis, Y. M. & Powell, H. K. J. (1966). *Journal of the Chemical Society, A.* 1015–8.

Curtis, N. F., Einstein, F. W. B. & Willis, A. C. (1984). *Inorganic Chemistry*, **23**, 3444–9.

Curtis, N. F. & House, D. A. (1961). *Chemistry and Industry*, 1708–9.

DelDonno, T. A. & Rosen, W. (1978). *Inorganic Chemistry*, **17**, 3714–6.

Drew, M. G. B., Esho, F. S. & Nelson, S. M. (1983). *Journal of The Chemical Society, Dalton Transactions*, 1653–9.

Drew, M. G. B., Esho, F. S., Lavery, A. & Nelson, S. M. (1984). *Journal of The Chemical Society, Dalton Transactions*, 545–6 and references therein.

Fernando, Q. & Wheatley, P. (1965). *Inorganic Chemistry*, **4**, 1726.

Green, M., Smith, J. & Tasker, P. A. (1971). *Inorganica Chimica Acta*, **5**, 17–24.

Hay, R. W., Lawrance, G. A. & Curtis, N. F. (1975). *Journal of The Chemical Society, Dalton Transactions*, 591–3.

Jackels, S. C., Farmery, K., Barefield, E. K., Rose, N. J. & Busch, D. H. (1972). *Inorganic Chemistry*, **11**, 2893–901.

Jäger, E.-G. (1964). *Zeitschrift Für Chemie*, **4**, 437.

Jäger, E.-G. (1969). *Zeitschrift Für Anorganische und Allgemeine Chemie*, **364**, 177–91 and references therein.

Johnston, D. L. & Horrocks, DeW. (1971). *Inorganic Chemistry*, **10**, 687–91.

Karbach, S., Löhr, W. & Vögtle, F. (1981). *Journal of Chemical Research (S)* 314–5.

Katovic, V., Taylor, L. T. & Busch, D. H. (1969). *Journal of the American Chemical Society*, **91**, 2122–3.

Kluiber, R. W. & Sasso, G. (1970). *Inorganica Chimica Acta*, **4**, 226–30.

Kyba, E. P. & Chou, S.-S. P. (1980). *Journal of the American Chemical Society*, **102**, 7012–4.

Kyba, E. P. & Chou, S.-S. P. (1981). *Journal of Organic Chemistry*, **46**, 860–3.

Lever, A. B. P. (1965). *Advances in Inorganic Chemistry & Radiochemistry*, **7**, 27–114.

Lindoy, L. F. & Busch, D. H. (1969). *Journal of the American Chemical Society*, **91**, 4690–3.

McGeachin, S. G. (1966). *Canadian Journal of Chemistry*, **44**, 2323–8.

Martin, J. G. & Cummings, S. C. (1973). *Inorganic Chemistry*, **12**, 1477–82.

Martin, J. G., Wei, R. M. C. & Cummings, S. C. (1972). *Inorganic Chemistry*, **11**, 475–9.

Marty, W. & Schwarzenbach, G. (1970). *Chimica*, **24**, 431–3.

Melson, G. A. & Busch, D. H. (1965). *Journal of the American Chemical Society*, **87**, 1706–10.

Melson, G. A. & Funke, L. A. (1984). *Inorganica Chimica Acta.*, **82**, 19–25.

Moser, F. & Thomas, A. (1963). *Phthalocyanine Compounds*, New York: Reinhold.

Nelson, S. M. (1980). *Pure and Applied Chemistry*, **52**, 461–76.

Ogawa, S. (1977). *Journal of The Chemical Society, Perkin Transactions* I, 214–16.

Ogawa, S., Yamaguchi, T. & Gotoh, N. (1974). *Journal of The Chemical Society, Perkin Transactions* I, 976–8.

Owston, P. G., Peters, R., Ramsammy, E., Tasker, P. A. & Trotter, J. (1980). *Journal of The Chemical Society, Chemical Communications*, 1218–20.

Richman, J. E. & Atkins, T. J. (1974). *Journal of the American Chemical Society*, **96**, 2268–70.

Rosen, W. & Busch, D. H. (1969). *Journal of the American Chemical Society*, **91**, 4694–7.

Scanlon, L. G., Sao, Y.-Y., Cummings, S. C., Toman, K. & Meek, D. W. (1980). *Journal of the American Chemical Society*, **102**, 6849–51.

Schrauzer, G. N. (1962). *Chemische Berichte*, **95**, 1438–45.

Stephens, F. S. & Vagg, R. S. (1977). *Acta Crystallographica*, **B33**, 3159–64.

Taylor, L. T. & Busch, D. H. (1967). *Journal of the American Chemical Society*, **89**, 5372–6.

Taylor, L. T., Urbach, F. L. & Busch, D. H. (1969). *Journal of the American Chemical Society*, **91**, 1072–5.

Taylor, L. T., Vergez, S. C. & Busch, D. H. (1966). *Journal of the American Chemical Society*, **88**, 3170–1.

Thompson, M. C. & Busch, D. H. (1964). *Journal of the American Chemical Society*, **86**, 3651–5.

Travis, K. & Busch, D. H. (1970). *Journal of The Chemical Society, Chemical Communications*, 1041–2.

248   *References*

Travis, K. & Busch, D. H. (1974). *Inorganic Chemistry*, **13**, 2591–8.
Uhlemann, E. & Plath, M. (1969). *Zeitschrift Für Chemie*, **9**, 234–5.
Umland, F. & Thierig, D. (1962). *Angewandte Chemie*, **74**, 388.
Welsh, W. A., Reynolds, G. J. & Henry, P. M. (1977). *Inorganic Chemistry*, **16**, 2558–61.

Chapter 3
Addison, A. W. (1976). *Inorganic and Nuclear Chemistry Letters*, **12**, 899–903.
Agnus, Y., Louis, R., Gisselbrecht, J.-P., & Weiss, P. (1984). *Journal of the American Chemical Society*, **106**, 93–102.
Albrecht-Gary, A. M., Saad, Z., Dietrich-Buchecker, C. O. & Sauvage, J. P. (1985). *Journal of the American Chemical Society*, **107**, 3205–9.
Barefield, E. K., Chueng, D., Van Derveer, D. G. & Wagner, F. (1981). *Journal of The Chemical Society, Chemical Communications*, 302–4.
Black, D. St. C., Vanderzalm, C. H. B. & Wong, L. C. H. (1979). *Australian Journal of Chemistry*, **32**, 2303–11.
Bond, A. M., Lawrance, G. A., Lay, P. A. & Sargeson, A. M. (1983). *Inorganic Chemistry*, **22**, 2010–21.
Boston, D. R. & Rose, N. J. (1968). *Journal of the American Chemical Society*, **90**, 6859–60.
Boston, D. R. & Rose, N. J. (1973). *Journal of the American Chemical Society*, **95**, 4163–8.
Boucher, H. A., Lawrance, G. A., Lay, P. A., Sargeson, A. M., Bond, A. M., Sangster, D. F. & Sullivan, J. C. (1983). *Journal of the American Chemical Society*, **105**, 4652–61 and references therein.
Buøen, S., Dale, J., Groth, P. & Krane, J. (1982). *Journal of The Chemical Society, Chemical Communications*, 1172–4.
Burk, P. L., Osborn, J. A., Youinou, M.-T., Agnus, Y., Louis R. & Weiss, R. (1981). *Journal of the American Chemical Society*, **103**, 1273–4.
Busch, D. H. (1980). *Pure and Applied Chemistry*, **52**, 2477–84.
Busch, D. H., Jackels, S. C., Callahan, R. C., Grzybowski, J. J., Zimmer, L. L., Kojima, M., Olszanski, D. J., Schammel, W. P., Stevens, J. C., Holter, K. A. & Mocak, J. (1981). *Inorganic Chemistry*, **20**, 2834–44.
Caron, A., Guilhelm, J., Riche, C., Pascard, C., Alpha, B., Lehn, J.-M. & Rodriguez-Ubis, J.-C. (1985). *Helvetica Chimica Acta*, **68**, 1577–82.
Cesario, M., Dietrich-Buchecker, C. O., Guilhem, J., Pascard, C. & Sauvage, J. P. (1985). *Journal of The Chemical Society, Chemical Communications*, 244–7.
Chang, C. K. (1977). *Journal of The Chemical Society, Chemical Communications*, 800–1.
Chang, C. A. & Rowland, M. E. (1983). *Inorganic Chemistry*, **22**, 3866–9.
Churchill, M. R. & Reis, A. H. (1973). *Journal of The Chemical Society, Dalton Transactions*, 1570–6 and references therein.
Collman, J. P., Elliott, C. M., Halbert, T. R. & Tovrog, B. S. (1977). *Proceedings of the National Academy of Science, USA*, **74**, 18–22.
Comarmond, J., Plumere, P., Lehn, J.-M., Agnus, Y., Louis, R., Weiss, R., Kahn, O. & Morgenstern-Badarau, I. (1982). *Journal of the American Chemical Society*, **104**, 6330–40.
Coughlin, P. K., Dewan, J. C., Lippard, S. J., Watanabe, E.-I. & Lehn, J.-M. (1979). *Journal of the American Chemical Society*, **101**, 265–6.
Coughlin, P. K. & Lippard, S. J. (1981). *Journal of the American Chemical Society*, **103**, 3228–9.

Coughlin, P. K., Lippard, S. J., Martin, A. E. & Bulkowski, J. E. (1980). *Journal of the American Chemical Society*, **102**, 7616–17.

Coughlin, P. K., Martin, A. E., Dewan, J. C., Watanabe, E.-I., Bulkowski, J. E., Lehn, J. M. & Lippard, S. J. (1984). *Inorganic Chemistry*, **23**, 1004–9.

Creaser, I. I., Harrowfield, J. MacB., Herlt, A. J., Sargeson, A. M., Springborg, J., Geue, R. J. & Snow, M. R. (1977). *Journal of the American Chemical Society*, **99**, 3181–2.

Creaser, I. I., Geue, R. J., Harrowfield, J., MacB., Herlt, A. J., Sargeson, A. M., Snow, M. R. & Springborg, J. (1982). *Journal of the American Chemical Society*, **104**, 6016–25.

Cunningham, J. A. & Sievers, R. E. (1973). *Journal of the American Chemical Society*, **95**, 7183–5.

Delgado, R. & da Silva, F. (1982). *Talanta*, **29**, 815–22.

Desreux, J. F. (1980). *Inorganic Chemistry*, **19**, 1319–24.

Deitrich-Buchecker, C. O., Sauvage, J.-P. & Kern, J.-M. (1984). *Journal of the American Chemical Society*, **106**, 3043–5.

Drew, M. G. B., Cairns, C., Lavery, A. & Nelson, S. M. (1980). *Journal of The Chemical Society, Chemical Communications*, 1122–4.

Drew, M. G. B., McCann, M. & Nelson, S. M. (1981). *Journal of The Chemical Society, Dalton Transactions*, 1868–78.

Drew, M. G. B., Rodgers, A., McCann, M. & Nelson, S. M. (1978). *Journal of The Chemical Society, Chemical Communications*, 415–6.

Durand, R. R., Bencosme, C. S., Collman, J. P. & Anson, F. C. (1983). *Journal of the American Chemical Society*, **105**, 2710–8 and references therein.

Fabbrizzi, L., Forlini, F., Perotti, A. & Seghi, B. (1984). *Inorganic Chemistry*, **23**, 807–13.

Fenton, D. E., Casellato, V., Vigato, P. A. & Vidali, M. (1982). *Inorganica Chimica Acta*, **62**, 57–66.

Fleischer, E. B., Sklar, L., Kendall-Torry, A., Tasker, P. A. & Taylor, F. B. (1973). *Inorganic and Nuclear Chemistry Letters*, **9**, 1061–7.

Freeman, G. M., Barefield, E. K. & Van Derveer, D. G. (1984). *Inorganic Chemistry*, **23**, 3092–103.

Gagne, R. R., Koval, C. A. & Smith, T. J. (1977). *Journal of the American Chemical Society*, **99**, 8367–8.

Gahan, L. R., Hambley, T. W., Sargeson, A. M. & Snow, M. R. (1982). *Inorganic Chemistry*, **21**, 2699–706.

Gahan, L. R., Lawrance, G. A. & Sargeson, A. M. (1982). *Australian Journal of Chemistry*, **35**, 1119–31.

Geue, R. J., Hambley, T. W., Harrowfield, J. MacB., Sargeson, A. M. & Snow, M. R. (1984). *Journal of the American Chemical Society*, **106**, 5478–88.

Geue, R. J., McCarthy, M. G. & Sargeson, A. M. (1984). *Journal of the American Chemical Society*, **106**, 8282–91.

Gunter, M. J., Mander, L. N., McLaughlin, G. M., Murray, K. S., Berry, K. J., Clark, P. E. & Buckingham, D. A. (1980). *Journal of the American Chemical Society*, **102**, 1470–3.

Hafliger, H. & Kaden, T. A. (1979). *Helvetica Chimica Acta*, **62**, 683–8.

Hammershøi, A. & Sargeson, A. M. (1983). *Inorganic Chemistry*, **22**, 3554–61.

Harris, W. R., Raymond, K. N. & Weitl, F. L. (1981). *Journal of the American Chemical Society*, **103**, 2667–75.

Harrowfield, J. MacB., Herlt, A. J., Lay, P. A., Sargeson, A. M., Bond, A. M., Mulac, W. A. & Sullivan, J. C. (1983). *Journal of the American Chemical Society*, **105**, 5503–5.

Hay, R. W. & Bembi, R. (1982). *Inorganica Chimica Acta*, **65**, L227–30.

Hay, R. W. & Clark, D. M. S. (1984). *Inorganica Chimica Acta*, **83**, L23–4.

Hediger, M. & Kaden, T. A. (1978). *Journal of The Chemical Society, Chemical Communications*, 14–15.

Herron, N., Chavan, M. Y. & Busch, D. H. (1984). *Journal of The Chemical Society, Dalton Transactions*, 1491–4.

Herron, N., Grzybowski, J. J., Matsumoto, N., Zimmer, L. L., Christoph, G. G. & Busch, D. H. (1982). *Journal of the American Chemical Society*, **104**, 1999–2007.

Herron, N., Schammel, W. P., Jackels, S. C., Grzybowski, J. J., Zimmer, L. L. & Busch, D. H. (1983). *Inorganic Chemistry*, **22**, 1433–40.

Hoskins, B. F., Robson, R. & Williams, G. A. (1976). *Inorganica Chimica Acta*, **16**, 121–33.

Hoskins, B. F. & Williams, G. A. (1975a). *Australian Journal of Chemistry*, **28**, 2607–14.

Hoskins, B. F. & Williams, G. A. (1975b). *Australian Journal of Chemistry*, **28**, 2593–605.

Jackels, S. C. & Rose, N. J. (1973). *Inorganic Chemistry*, **12**, 1232–7 and references therein.

Kagan, N. E., Mauzerall, D. & Merrifield, R. B. (1977) *Journal of the American Chemical Society*, **99**, 5484–6.

Kasprzyk, S. P. & Wilkins, R. G. (1982). *Inorganic Chemistry,* **21**, 3349–52.

Keypour, H. & Stotter, D. A. (1979). *Inorganica Chimica Acta*, **33**, L149–50.

Korybut-Daszkiewicz, B., Kojima, M., Cameron, J. H., Herron, N., Chavan, M. Y., Jircitano, A. J., Coltrain, B. K., Neer, G. L., Alcock, N. W. & Busch, D. H. (1984). *Inorganic Chemistry*, **23**, 903–14.

Kwik, W.-L., Herron, N., Takeuchi, K. & Busch, D. H. (1983). *Journal of The Chemical Society, Chemical Communications*, 409–11 and references therein.

Landrum, J. T., Grimmett, D., Haller, K. J., Scheidt, W. R. & Reed, C. A. (1981). *Journal of the American Chemical Society*, **103**, 2640–50.

Larsen, E., La Mar, G. N., Wagner, B. E., Parks, J. E. & Holm, R. H. (1972). *Inorganic Chemistry*, **11**, 2652–68.

Lehn, J. M. (1978). *Accounts of Chemical Research*, **11**, 49–57.

Lehn, J. M. (1980). *Pure and Applied Chemistry*, **52**, 2441–59.

Lehn, J. M., Pine, S. H., Watanabe, E. & Willard, A. K. (1977). *Journal of the American Chemical Society*, **99**, 6766–8.

Lotz, T. J. & Kaden, T. A. (1978). *Helvetica Chimica Acta*, **61**, 1376–87.

Louis, R., Agnus, Y. & Weiss, R. (1978). *Journal of the American Chemical Society*, **100**, 3604–5.

Madeyski, C. M., Michael, J. P., & Hancock, R. D. (1984). *Inorganic Chemistry*, **23**, 1487–9.

Motekaitis, R. J., Martell, A. E., Lecomte, J.-P. & Lehn, J.-M. (1983). *Inorganic Chemistry*, **22**, 609–14.

Motekaitis, R. J., Martell, A. E., Lehn, J.-M. & Watanabe, E.-I. (1982). *Inorganic Chemistry*, **21**, 4253–7.

Murase, I., Hamada, K. & Kida, S. (1981). *Inorganica Chimica Acta*, **54**, L171–3.

Nelson, S. M. (1982). *Inorganica Chimica Acta*, **62**, 39–50.

Nelson, S. M., Esho, F., Lavery, A. & Drew, M. G. B. (1983). *Journal of the American Chemical Society*, **105**, 5693–5 and references therein.

Parks, J. E., Wagner, B. E. & Holm, R. H. (1971). *Inorganic Chemistry*, **10**, 2472–8.

Pilkington, N. H. & Robson, R. (1970). *Australian Journal of Chemistry*, **23**, 2225–36.

Richardson, J. S., Thomas, K. A., Rubin, B. H. & Richardson, D. C. (1975). *Proceedings of the National Academy of Science, USA*, **72**, 1349–53.

Sargeson, A. M. (1979). *Chemistry in Britain*, **15**, 23–7.

Sargeson, A. M. (1984). *Pure and Applied Chemistry*, **56**, 1603–19.

Spirlet, M.-R., Rebizant, J., Desreux, J. F. & Loncin, M.-F. (1984). *Inorganic Chemistry*, 23, 359–63.

Stetter, H. & Frank, W. (1976). *Angewandte Chemie, International Edition in English*, **15**, 686.

Takahashi, M. & Takamoto, S. (1977). *Bulletin of the Chemical Society of Japan*, **50**, 3413–14.

Takeuchi, K. J., Busch, D. H. & Alcock, N. (1981). *Journal of the American Chemical Society*, **103**, 2421–2 and references therein.

Takeuchi, K. J. & Busch, D. H. (1983). *Journal of the American Chemical Society*, **105**, 6812–6.

Travis, K. & Busch, D. H. (1970). *Journal of The Chemical Society, Chemical Communications*, 1041–2.

Uechi, T., Ueda, I., Tazaki, M., Takagi, M. & Ueno, K. (1982). *Acta Crystallographica*, **B38**, 433–6.

van der Merwe, M. J., Boeyens, J. C. A. & Hancock, R. D. (1983). *Inorganic Chemistry*, **22**, 3489–90.

Wainwright, K. P. (1980). *Journal of The Chemical Society, Dalton Transactions*, 2117–20.

Wainwright, K. P. (1983). *Journal of The Chemical Society, Dalton Transactions*, 1149–52.

Weitl, F. L., Raymond, K. N., Smith, W. L. & Howard, T. R. (1978). *Journal of the American Chemical Society*, **100**, 1170–2.

Weitl, F. L. & Raymond, K. N. (1979). *Journal of the American Chemical Society*, **101**, 2728–31.

Wieghardt, K., Bossek, U., Chaudhuri, P., Herrmann, W., Menke, B. C. & Weiss, J. (1982). *Inorganic Chemistry*, **21**, 4308–14.

Williams, G. A. & Robson, R. (1981). *Australian Journal of Chemistry*, **34**, 65–79.

Zakrzewski, G. A., Ghilardi, C. A. & Lingafelter, E. C. (1971). *Journal of the American Chemical Society*, **93**, 4411–15.

Chapter 4

Almasio, M. C., Arnaud-Neu, F. & Schwing-Weill, M. J. (1983). *Helvetica Chimica Acta*, **66**, 1296–306.

Anderegg, G. (1981). *Helvetica Chimica Acta*, **64**, 1790–5.

Arnaud-Neu, F., Spiess, B. & Schwing-Weill, M. J. (1977) *Helvetica Chimica Acta*, **60**, 2633–43.

Arnaud-Neu, F., Spiess, B. & Schwing-Weill, M. J. (1982). *Journal of the American Chemical Society*, **104**, 5641–5.

Blasius, E., Janzen, K.-P., Klein, W., Klotz, H., Nguyen, V. B., Nguyen-Tien, T., Pfeiffer, R., Scholten, G., Simon, H., Stockemer, H. & Toussaint, A. (1980). *Journal of Chromatography*, **201**, 147–66.

Blasius, E. & Janzen, K.-P. (1981). *Topics in Current Chemistry*, **98**, 163–97.

Blasius, E. & Janzen, K.-P. (1982). *Pure and Applied Chemistry*, **54**, 2115–28.

Bradshaw, J. S., Reeder, R. A., Thompson, M. D., Flanders, E. D., Carruth, R. L., Izatt, R. M. & Christensen, J. J. (1976). *Journal of Organic Chemistry*, **41**, 134–6 and references therein.

Bright, D. & Truter, M. R. (1970). *Nature*, **225**, 176–7.

Buhleier, E., Wehner, W. & Vogtle, F. (1978). *Chemische Berichte*, **111**, 200–4.

Burns, J. H. & Baes, C. F. (1981). *Inorganic Chemistry*, **20**, 616–19.

Bush, M. A. & Truter, M. R. (1972). *Journal of The Chemical Society, Perkin Transactions II*, 341–4.

Cahen, Y. M., Dye, J. L. & Popov, A. I. (1975). *Journal of Physical Chemistry*, **79**, 1289–91.

Cahen, Y. M., Dye, J. L. & Popov, A. I. (1975). *Journal of Physical Chemistry*, **79**, 1292–5.

Cheney, J. & Lehn, J.-M. (1972). *Journal of The Chemical Society, Chemical Communications*, 487–9.

Cho, I. & Chang, S.-K. (1980). *Bulletin of the Korean Chemical Society*, **1**, 145–6.

Cinquini, M., Colonna, S., Molinari, H., Montanari, F. & Tundo, P. (1976). *Journal of The Chemical Society, Chemical Communications*, 394–6.

Cook, F. L., Caruso, T. C., Byrne, M. P., Bowers, C. W., Speck, D. H. & Liotta, C. L. (1974). *Tetrahedron Letters*, 4029–32.

Coxon, A. C. & Stoddart, J. F. (1977). *Journal of The Chemical Society, Perkin Transactions* I, 767–85.

Czugler, M. & Weber, E. (1981). *Journal of The Chemical Society, Chemical Communications*, 472–3.

Dale, J. & Daasvatn, K. (1976). *Journal of The Chemical Society, Chemical Communications*, 295–6.

Davidson, R. B., Izatt, R. M., Christensen, J. J., Schultz, R. A., Dishong, D. M. & Gokel, G. W. (1984), *Journal of Organic Chemistry*, **49**, 5080–4.

Davies, J. E. D., Kemula, W., Powell, H. M. & Smith, N. O. (1983). *Journal of Inclusion Phenomena*, **1**, 3–44.

Davydova, S. P., Baravanov, V. A., Apymova, N. V. & Prata, N. A. (1975). *Izvestiya Akademii Nauk SSSR, Ser. Khim.*, 1441–3.

Desvergne, J. P. & Bouas-Laurent, H. (1978). *Journal of The Chemical Society, Chemical Communications*, 403–4.

Dietrich, B., Lehn, J.-M. & Sauvage, J. P. (1969). *Tetrahedron Letters*, **34**, 2885–8.

Dietrich, B., Lehn, J.-M. & Sauvage, J. P. (1970). *Journal of The Chemical Society, Chemical Communications*, 1055–6.

Dietrich, B., Lehn, J.-M. & Sauvage, J. P. (1973a). *Tetrahedron*, **29**, 1647–58.

Dietrich, B., Lehn, J.-M. & Sauvage, J. P. (1973b). *Journal of The Chemical Society, Chemical Communications*, 15–6.

Dietrich, B., Lehn, J.-M., Sauvage, J. P. & Blanzat, J. (1973). *Tetrahedron*, **29**, 1629–45.

Dix, J. P. & Vogtle, F. (1981). *Chemische Berichte*, **114**, 638–51 and references therein.

Dobler, M., Dunitz, J. D. & Seiler, P. (1974). *Acta Crystallographica*, **B30**, 2741–3.

Dye, J. L. (1984). *Progress in Inorganic Chemistry*, **32**, 327–441.

Dye, J. L. & Ellaboudy, A. (1984), *Chemistry in Britain*, 210–15.

Dye, J. L., Andrews, C. W. & Ceraso, J. M. (1975). *Journal of Physical Chemistry*, **79**, 3076–9.

Dye, J. L., Ceraso, J. M., Lok, M. T., Barnett, B. L. & Tehan, F. J. (1974). *Journal of the American Chemical Society*, **96**, 608–9.

Ellaboudy, A., Dye, J. L. & Smith, P. B. (1983). *Journal of the American Chemical Society*, **105**, 6490–1.

Fenton, D. E., Mercer, M., Poonia, N. S. & Truter, M. R. (1972). *Journal of The Chemical Society, Chemical Communications*, 66–7.

Fischer, J., Mellinger, M. & Weiss, R. (1977). *Inorganica Chimica Acta*, **21**, 259–63.

Frensdorff, H. K. (1971). *Journal of the American Chemical Society*, **93**, 600–6.

Gokel, G. W. & Korzeniowski, S. H. (1982). 'Macrocyclic Polyether Syntheses', Vol. 13 in

Reactivity and Structure Concepts in Organic Chemistry, eds. K. Hafner *et al.* Springer-Verlag, Berlin.

Greene, R. N. (1972). *Tetrahedron Letters*, 1793–6.

Handyside, T. M., Lockhart, J. C., McDonnell, M. B. & Rao, P. V. S. (1982). *Journal of The Chemical Society, Dalton Transactions*, 2331–6.

Herbert, J. A. & Truter, M. R. (1980). *Journal of The Chemical Society, Perkin Transactions* II, 1253–8.

Herceg, M. & Weiss, R. (1970). *Inorganic Nuclear Chemical Letters*, **6**, 435–7.

Hilgenfeld, R. & Saenger, W. (1982). *Topics in Current Chemistry*, **101**, 1–82.

Hiraoka, M. (1982). *Crown Compounds: their Characteristics and Applications*. Elsevier, Amsterdam.

Iwachido, T., Sadakane, A. & Toei, K. (1978). *Bulletin of the Chemical Society of Japan*, **51**, 629–30.

Izatt, R. M., Bradshaw, J. S., Nielsen, S. A., Lamb, J. D., Christensen, J. J. & Sen, D. (1985). *Chemical Reviews*, **85**, 271–339.

Izatt, R. M., Lamb, J. D., Maas, G. E., Asay, R. E., Bradshaw, J. S. & Christensen, J. J. (1977a). *Journal of the American Chemical Society*, **99**, 2365–6.

Izatt, R. M., Lamb, J. D., Assay, R. E., Maas, G. E., Bradshaw, J. S., Christensen, J. J. & Moore, S. S. (1977b). *Journal of the American Chemical Society*, **99**, 6134–6.

Kaifer, A., Durst, H. D., Echegoyen, L., Dishong, D. M., Schultz, R. A. & Gokel, G. W. (1982). *Journal of Organic Chemistry*, **47**, 3195–7.

Kaneda, T., Sugihara, K., Kamiya, H. & Misumi, S. (1981). *Tetrahedron Letters*, **22**, 4407–8.

Kimura, T., Iwashima, K. & Hamada, T. (1979). *Analytical Chemistry*, **51**, 1113–16.

Kirch, M. & Lehn, J.-M. (1975). *Angewandte Chemie, International Edition in English*, **14**, 555–6.

Knipe, A. C. (1976). *Journal of Chemical Education*, **53**, 618–22.

Kobuke, Y., Hanji, K., Horiguchi, K., Asada, M., Nakayama, Y. & Furukawa, J. (1976). *Journal of the American Chemical Society*, **98**, 7414–9.

Koenig, K. E., Helgeson, R. C. & Cram, D. J. (1976). *Journal of the American Chemical Society*, **98**, 4018–20.

Koltoff, I. M. (1979). *Analytical Chemistry*, **51**, 1R–22R.

Kresge, A. J. (1975). *Accounts of Chemical Research*, **8**, 354–60.

Lamb, J. D., Izatt, R. M., Christensen, J. J. & Eatough, D. J. (1979). In *Coordination Chemistry of Macrocyclic Compounds*, ed. G. A. Melson, pp. 145–217. Plenum Press, New York.

Lehn, J.-M. (1978). *Accounts of Chemical Research*, **11**, 49–56.

Lehn, J.-M. (1980). *Pure & Applied Chemistry*, **52**, 2441–59.

Lehn, J.-M. & Montavon, F. (1978). *Helvetica Chimica Acta*, **61**, 67–82.

Lehn, J.-M. & Sauvage, J. P. (1975). *Journal of the American Chemical Society*, **97**, 6700–7.

Lehn, J.-M. & Simon, J. (1977). *Helvetica Chimica Acta*, **60**, 141–51.

Lehn, J.-M., Simon, J. & Wagner, J. (1973). *Angewandte Chemie, International Edition in English*, **7**, 578–9.

Lehn, J.-M. & Stubbs, M. E. (1974). *Journal of the American Chemical Society*, **96**, 4011–12.

Liotta, C. L. & Harris, H. P. (1974). *Journal of the American Chemical Society*, **96**, 2250–2.

Lockhart, J. C., Robson, A. C., Thompson, M. E., Furtado, D., Kaura, C. K. & Allan, A. R. (1973). *Journal of The Chemical Society, Perkin Transactions* I, 577–81.

Lok, M. T., Tehan, F. J. & Dye, J. L. (1972). *Journal of Physical Chemistry*, **76**, 2975–81.
Luboch, E., Cygan, A. & Biernat, J. F. (1983). *Inorganica Chimica Acta*, **68**, 201–4.
Mallinson, P. R. & Truter, M. R. (1972). *Journal of The Chemical Society, Perkin Transactions* II, 1818–23.
Mandolini, L. & Masci, B. (1984). *Journal of the American Chemical Society*, **106**, 168–74.
Masuyama, A., Nakatsuji, Y., Ikeda, I. & Okahara, M. (1981). *Tetrahedron Letters*, **22**, 4665–8.
Matalon, S., Golden, S. & Ottolenghi, M. (1969). *Journal of Physical Chemistry*, **73**, 3098–101.
Matsuda, T. & Koida, K. (1973). *Bulletin of the Chemical Society of Japan*, **46**, 2259–60.
Mattice, W. L. & Newkome, G. R. (1982). *Journal of the American Chemical Society*, **104**, 5942–4.
Metz, B., Moras, D. & Weiss, R. (1976). *Journal of The Chemical Society, Perkin Transactions* II, 423–9.
Montanari, F., Landini, D. & Rolla, F. (1982). *Topics in Current Chemistry*, **101**, 147–200.
Moras, D., Metz, B. & Weiss, R. (1973). *Acta Crystallographica*, **B29**, 388–95.
Nakamura, H., Nishida, H., Takagi, M. & Ueno, K. (1982). *Analytica Chimica Acta*, **139**, 219–27.
Nakamura, H., Sakka, H., Takagi, M. & Ueno, K. (1981). *Chemistry Letters*, 1305–6.
Newcomb, M. & Cram, D. J. (1975). *Journal of the American Chemical Society*, **97**, 1257–9.
Newkome, G. R., Sauer, J. D., Roper, J. M. & Hager, D. C. (1977). *Chemical Reviews*, **77**, 513–97.
Newkome, G. R., Majestic, V., Fronczek, F. & Atwood, J. L. (1979). *Journal of the American Chemical Society*, **101**, 1047–8.
Nishida, H., Tazaki, M., Takagi, M. & Ueno, K. (1981), *Mikrochimica Acta*, **1**, 281–7.
Owen, J. D. (1983). *Journal of The Chemical Society, Perkin Transactions* II, 407–15.
Parsons, D. G. (1978). *Journal of The Chemical Society, Perkin Transactions* I, 451–5.
Pedersen, C. J. (1967). *Journal of the American Chemical Society*, **89**, 7017–36.
Pedersen, C. J. (1971). *Journal of Organic Chemistry*, **36**, 254–7.
Regen, S. L. (1979). *Angewandte Chemie, International Edition in English*, **18**, 421–9.
Reinhoudt, D. N. & Gray, R. T. (1975). *Tetrahedron Letters*, **25**, 2105–8.
Sam, D. J. & Simmons, H. F. (1972). *Journal of the American Chemical Society*, **94**, 4024–5.
Sam, D. J. & Simmons, H. F. (1974). *Journal of the American Chemical Society*, **96**, 2252–3.
Schultz, R. A., Dishong, D. M. & Gokel, G. M. (1982). *Journal of the American Chemical Society*, **104**, 625–6.
Seiler, P., Dobler, M. & Dunitz, J. D. (1974). *Acta Crystallographica*, **B30**, 2744–5.
Shiga, N., Takagi, M. & Ueno, K. (1980). *Chemistry Letters*, 1021–2.
Shinkai, S., Ogawa, T., Kusano, Y., Manabe, O., Kikukawa, K., Goto, T. & Matsuda, T. (1982). *Journal of the American Chemical Society*, **104**, 1960–7.
Shinkai, S., Minami, T., Kusano, Y. & Manabe, O. (1983). *Journal of the American Chemical Society*, **105**, 1851–6.
Shinkai, S., Nakaji, T., Nishida, Y., Ogawa, T. & Manabe, O. (1980). *Journal of the American Chemical Society*, **102**, 5860–5.
Shinkai, S. & Manabe, O. (1984). *Topics in Current Chemistry*, **121**, 67–104.
Shinkai, S., Shigematsu, K., Sato, M. & Manabe, O. (1982). *Journal of The Chemical Society, Perkin Transactions* I, 2735–9.
Shoham, G., Lipscomb, W. N. & Olsher, U. (1983a). *Journal of The Chemical Society, Chemical Communications*, **105**, 208–9.

Shoham, G., Lipscomb, W. N. & Olsher, U. (1983b). *Journal of the American Chemical Society*, **105**, 1247–52.

Takagi, M. & Ueno, K. (1984). *Topics in Current Chemistry*, **121**, 39–65.

Takeda, Y. & Kato, H. (1979). *Bulletin of the Chemical Society of Japan*, **52**, 1027–30.

Tehan, F. J., Barnett, B. L. & Dye, J. L. (1974). *Journal of the American Chemical Society*, **96**, 7203–8.

Timko, J. M. & Cram, D. J. (1974). *Journal of the American Chemical Society*, **96**, 7159–60.

Timko, J. M., Helgeson, R. C., Newcomb, M., Gokel, G. W. & Cram, D. J. (1974). *Journal of the American Chemical Society*, **96**, 7097–9.

Truter, M. R. (1973). *Chemistry in Britain*, 203–7.

Vogtle, F. (1980). *Pure and Applied Chemistry*, **52**, 2405–16.

Vogtle, F. & Weber, E. (1974). *Angewandte Chemie, International Edition in English*, **13**, 149–50.

Walba, D. M., Richards, R. M. & Haltiwanger, R. C. (1982). *Journal of the American Chemical Society*, **104**, 3219–21.

Weber, E. (1979). *Angewandte Chemie, International Edition in English*, **18**, 219–20.

Weber, E. (1982). *Journal of Organic Chemistry*, **47**, 3478–86.

Weber, E. & Vogtle, F. (1980). *Inorganica Chimica Acta*, **45**, L65–7.

Weber, E. & Vogtle, F. (1981). *Topics in Current Chemistry*, **98**, 1–41.

Weber, W. P. & Gokel, G. W. (1977). *Phase Transfer Catalysis in Organic Synthesis*. Springer-Verlag, Berlin.

Wong, K. H., Konizer, G. & Smid, J. (1970). *Journal of the American Chemical Society*, **92**, 666–70.

Yee, E. L., Gansow, O. A. & Weaver, M. J. (1980). *Journal of the American Chemical Society*, **102**, 2278–5.

Chapter 5

Arduini, A., Pochini, A., Reverberi, S. & Ungaro, R. (1984). *Journal of The Chemical Society, Chemical Communications*, 981–2.

Behr, J. P. & Lehn, J. M. (1976). *Journal of the American Chemical Society*, **98**, 1743–7.

Behr, J. P., Lehn, J. M. & Vierling, P. (1976). *Journal of the Chemical Society, Chemical Communications*, 621–3.

Bender, M. L. & Komiyama, M. (1978). *Cyclodextrin Chemistry*. Springer, Berlin.

Bergeron, R. J. (1977). *Journal of Chemical Education*, 204–7.

Breslow, R. (1971). *Advances in Chemistry Series*, **100**, 21–43.

Breslow, R. (1982). *Science*, **218**, 532–7.

Breslow, R. (1983). *Chemistry in Britain*, 126–31.

Breslow, R. & Campbell, P. (1969). *Journal of the American Chemical Society*, **91**, 3085.

Breslow, R. & Campbell, P. (1971). *Bioorganic Chemistry*, **1**, 140–56.

Breslow, R., Hammond, M. & Lauer, M. (1980). *Journal of the American Chemical Society*, **102**, 421–2 and references therein.

Breslow, R. & Overman, L. E. (1970). *Journal of the American Chemical Society*, **92**, 1075–7.

Canceill, J., Lacombe, L. & Collet, A. (1986). *Journal of the American Chemical Society*, **108**, 4230–2.

Colquhoun, H. M., Lewis, D. F., Stoddart, J. F. & Williams, D. J. (1983). *Journal of The Chemical Society, Dalton Transactions*, 607–13.

Cram, D. J. (1983). *Science*, **219**, 1177–83.

Cram, D. J. & Cram, J. M. (1978). *Accounts of Chemical Research*, **11**, 8–14.

Cram, D. J., Dicker, I. B., Lein, G. M., Knobler, C. B. & Trueblood, K. N. (1982). *Journal of the American Chemical Society*, **104**, 6827–8.

Cram, D. J. & Ho, S. P. (1986). *Journal of the American Chemical Society*, **108**, 2998–3005.

Cram, D. J., Ho, S. P., Knobler, C. B., Maverick, E. & Trueblood, K. N. (1986). *Journal of the American Chemical Society*, **108**, 2989–98.

Cram, D. J., Kaneda, T., Helgeson, R. C. & Lein, G. M. (1979). *Journal of the American Chemical Society*, **101**, 6752–4.

Cram, D. J., Katz, H. E. & Dicker, I. B. (1984). *Journal of the American Chemical Society*, **106**, 4987–5000.

Cram, D. J., Lein, G. M., Kaneda, T., Helgeson, R. C., Knobler, C. B., Maverick, E. & Trueblood, K. N. (1981). *Journal of the American Chemical Society*, **103**, 6228–32.

Cram, D. J. & Trueblood, K. N. (1981). *Topics in Current Chemistry*, **98**, 43–106.

Cramer, F., Saenger, W. & Spatz, H.-Ch. (1967). *Journal of the American Chemical Society*, **98**, 14–20.

Davidson, R. B., Bradshaw, J. S., Jones, B. A., Dalley, N. K., Christensen, J. J., Izatt, R. M., Morin, F. G. & Grant, D. M. (1984). *The Journal of Organic Chemistry*, **49**, 353–7.

de Jong, F., Reinhoudt, D. N. & Smit, C. J. (1976). *Tetrahedron Letters*, **16**, 1371–4.

Dietrich, B., Guilhem, J., Lehn, J. M., Pascard, C. & Sonveaux, E. (1984). *Helvetica Chimica Acta*, **67**, 91–104.

Faust, G. & Pallas, M. (1960). *Journal fur Praktische Chemie*, **11**, 146–52.

Fujita, K., Ejima, S. & Imoto, T. (1984). *Journal of the Chemical Society, Chemical Communications*, 1277–9 and references therein.

Gabard, J. & Collet, A. (1981). *Journal of the Chemical Society, Chemical Communications*, 1137–9.

Girodeau, J. M., Lehn, J. M. & Sauvage, J. P. (1975). *Angewandte Chemie, International Edition in English*, **14**, 654.

Gokel, G. W., Cram, D. J., Liotta, C. L., Harris, H. P. & Cook, F. L. (1974). *Journal of Organic Chemistry*, **39**, 2445–6.

Goldberg, I. (1975a). *Acta Crystallographica*, **B31**, 2592–600.

Goldberg, I. (1975b). *Acta Crystallographica*, **B31**, 754–62.

Goldberg, I. (1977). *Acta Crystallographica*, **B33**, 472–9.

Goldberg, I. (1978). *Acta Crystallographica*, **B34**, 3387–90.

Graf, E. & Lehn, J. M. (1975). *Journal of the American Chemical Society*, **97**, 5022–4.

Graf, E. & Lehn, J. M. (1976). *Journal of the American Chemical Society*, **98**, 6403–5.

Griffiths, D. W. & Bender, M. L. (1973). In *Advances in Catalysis, Vol. 23*, ed. H. Pines & P. B. Weisz. Academic Press, New York.

Gutsche, D. B., Dhawan, B., No, K. H. & Muthukrishnan, R. (1981). *Journal of the American Chemical Society*, **103**, 3782–92.

Gutsche, D. C. & Levine, J. A. (1982). *Journal of the American Chemical Society*, **104**, 2652–3.

Helgeson, R. C., Mazaleyrat, J.-P. & Cram, D. J. (1981). *Journal of the American Chemical Society*, **103**, 3929–31.

Helgeson, R. C., Tarnowski, T. L. & Cram, D. J. (1979). *Journal of Organic Chemistry*, **44**, 2538–50.

Helgeson, R. C., Timoko, J. M., Moreau, P., Peacock, S. C., Mayer, J. M. & Cram, D. J. (1974). *Journal of the American Chemical Society*, **96**, 6762–3.

Hennrich, N. & Cramer, F. (1965). *Journal of the American Chemical Society*, **87**, 1121–6.

Katz, H. E. & Cram, D. J. (1984). *Journal of the American Chemical Society*, **106**, 4977–87.

Kintzinger, J. P., Lehn, J. M., Kauffmann, E., Dye, J. L. & Popov, A. I. (1983). *Journal of the American Chemical Society*, **105**, 7549–53.

Kyba, E. P., Helgeson, R. C., Madan, K., Gokel, G. W., Tarnowski, T. L., Moore, S. S. & Cram, D. J. (1977). *Journal of the American Chemical Society*, **99**, 2564–71.

Lehn, J. M., Simon, J. & Moradpour, A. (1978). *Helvetica Chimica Acta*, **61**, 2407–18.

Lehn, J. M., Pine, S. H., Watanabe, E.-I. & Willard, A. K. (1977). *Journal of the American Chemical Society*, **99**, 6766–8.

Lein, G. M. & Cram, D. J. (1982). *Journal of The Chemical Society, Chemical Communications*, 302–4.

Meade, T. J. & Busch, D. H. (1985). *Progress in Inorganic Chemistry*, **33**, 59–126.

Metz, B., Rozalky, J. M. & Weiss, R. (1976). *Journal of The Chemical Society, Chemical Communications*, 533–4.

Moran, J. R., Karbach, S. & Cram, D. J. (1982). *Journal of the American Chemical Society*, **104**, 5826–8.

Motekaitis, R. J., Martell, A. E., Lehn, J. M. & Watanabe, E. (1982). *Inorganic Chemistry*, **21**, 4253–7.

Nagano, O., Kobayashi, A. & Sasaki, Y. (1978). *Bulletin of the Chemical Society, Japan*, **51**, 790–3.

Newcomb, M., Timko, J. M., Walba, D. M. & Cram, D. J. (1977). *Journal of the American Chemical Society*, **99**, 6392–8.

Newcomb, M., Moore, S. S. & Cram, D. J. (1977). *Journal of the American Chemical Society*, **99**, 6405–10.

Nolte, R. J. M. & Cram, D. J. (1984). *Journal of the American Chemical Society*, **106**, 1416–20.

Odashima, K., Itai, A., Iitaka, Y. & Koga, K. (1980). *Journal of the American Chemical Society*, **102**, 2504–5.

Park, C. H. & Simmons, H. E. (1968). *Journal of the American Chemical Society*, **90**, 2431–2 and references therein.

Peacock, S. C. & Cram, D. J. (1976). *Journal of The Chemical Society, Chemical Communications*, 282–4.

Pedersen, C. J. (1967). *Journal of the American Chemical Society*, **89**, 7017–36.

Pedersen, C. J. (1971). *Journal of Organic Chemistry*, **36**, 1690–3.

Saenger, W. (1980). *Angewandte Chemie, International Edition in English*, **19**, 344–62.

Schmidtchen, F. P. (1980). *Chemische Berichte*, **113**, 864–74.

Shchori, E. & Jagur-Grodzinski, J. (1972). *Israel Journal of Chemistry*, **10**, 935–40.

Simmons, H. E. & Park, C. H. (1968). *Journal of the American Chemical Society*, **90**, 2428–9 and references therein.

Stetter, H. & Roos, E.-E. (1955). *Chemische Berichte*, **88**, 1390–5.

Stoddart, J. F. (1979). *Chemical Society Reviews*, **8**, 85–142.

Sutherland, I. O. (1986). *Chemical Society Reviews*, **15**, 63–91.

Szejtli, J. (Ed.) (1982). *Cyclodextrins*. Reidel, Dordrecht, the Netherlands.

Tabushi, I. (1984). *Tetrahedron*, **40**, 269–92.

Tabushi, I., Kimura, Y. & Yamamura, K. (1982). In *Chemical Approaches to Understanding Enzyme Catalysis*, ed. G. S. Green, Y. Ashami & D. Chipman. Elsevier, Amsterdam.

Tabushi, I., Kiyosuke, Y., Sugimoto, T. & Yamamura, K. (1978). *Journal of the American Chemical Society*, **100**, 916–9.

Tabushi, I., Kuroda, Y. & Kimura, Y. (1976). *Tetrahedron Letters*, 3327–30.

Tabushi, I., Kuroda, Y. & Mochizuki, J. (1980). *Journal of the American Chemical Society*, **102**, 1152–3 and references therein.

Tabushi, I., Kuroda, Y. & Shimokawa, K. (1979). *Journal of the American Chemical Society*, **101**, 1614–5.

Tabushi, I., Sasaki, H. & Kuroda, Y. (1976). *Journal of the American Chemical Society*, **98**, 5727–8.

Tabushi, I., Shimizu, N., Sugimoto, T., Shiouzuka, M. & Yamamura, K. (1977). *Journal of the American Chemical Society*, **99**, 7100–2.

Tabushi, I. & Kuroda, Y. (1984). *Journal of the American Chemical Society*, **106**, 4580–4.

Timko, J. M., Moore, S. S., Walba, D. M., Hiberty, P. C. & Cram, D. J. (1977). *Journal of the American Chemical Society*, **99**, 4207–19.

Trainor, G. L. & Breslow, R. (1981). *Journal of the American Chemical Society*, **103**, 154–8.

Trueblood, K. N., Knobler, C. B., Lawrence, D. S. & Stevens, R. V. (1982). *Journal of the American Chemical Society*, **104**, 1355–62.

Van Etten, R. L., Clowes, G. A., Sebastian, J. F. & Bender, M. L. (1967). *Journal of the American Chemical Society*, **89**, 3253–62 and references therein.

Villiers, A. (1891). *Comptes Rendus Hebdomadaires des Seances de l'Academic des Sciences, Paris*, **112**, 536.

Vogtle, F., Sieger, H. & Muller, W. M. (1981). *Topics in Current Chemistry*, **98**, 107–61.

Chapter 6

Abraham, M. H., de Namor, A. F. D. & Lee, W. H. (1977). *Journal of The Chemical Society, Chemical Communications*, 893–4.

Anderegg, G. (1975). *Helvetica Chimica Acta*, **58**, 1218–25.

Anderegg, G., Ekstrom, A., Lindoy, L. F. & Smith, R. J. (1980). *Journal of the American Chemical Society*, **102**, 2670–4.

Arnaud-Neu, F., Schwing-Weill, M.-J., Louis, R. & Weiss, R. (1979). *Inorganic Chemistry*, **11**, 2956–61.

Bianchi, A., Bologni, L., Dapporto, P., Micheloni, M. & Paoletti, P. (1984). *Inorganic Chemistry*, **23**, 1201–5.

Cabbiness, D. K. & Margerum, D. W. (1969). *Journal of the American Chemical Society*, **91**, 6540–2.

Cabbiness, D. K. & Margerum, D. W. (1970). *Journal of the American Chemical Society*, **92**, 2151–3.

Clay, R. M., Corr, S., Micheloni, M. & Paoletti, P. (1985). *Inorganic Chemistry*, **24**, 3330–6.

Clay, R. M., McCormac, H., Micheloni, M. & Paoletti, P. (1982). *Inorganic Chemistry*, **21**, 2494–6.

Clay, R. M., Micheloni, M., Paoletti, P. & Steele, W. V. (1979). *Journal of the American Chemical Society*, **101**, 4119–22.

Diaddario, L. L., Zimmer, L. L., Jones, T. E., Sokol, L. S. W. L., Cruz, R. B., Yee, E. L., Ochrymowycz, L. A. & Rorabacher, D. B. (1979). *Journal of the American Chemical Society*, **101**, 3511–20.

Dietrich, B., Lehn, J.-M., Sauvage, J.-P. (1973). *Journal of The Chemical Society, Chemical Communications*, 15–7.

Frensdorff, H. K. (1971). *Journal of the American Chemical Society*, **93**, 600–6.

Haymore, B. L., Lamb, J. D., Izatt, R. M. & Christensen, J. J. (1982). *Inorganic Chemistry*, **21**, 1598–1602.

Henrick, K., Lindoy, L. F., McPartlin, M., Tasker, P. A. & Wood, M. P. (1984). *Journal of the American Chemical Society*, **106**, 1641–5.

Henrick, K., Tasker, P. A. & Lindoy, L. F. (1985). *Progress in Inorganic Chemistry*, **33**, 1–58.

Hinz, F. P. & Margerum, D. W. (1974). *Inorganic Chemistry*, **13**, 2941–9.

Izatt, R. M., Bradshaw, J. S., Nielsen, S. A., Lamb, J. D., Christensen, J. J. & Sen, D. (1985). *Chemical Reviews*, **85**, 271–339.

Kauffmann, E., Lehn, J.-M. & Sauvage, J.-P. (1976). *Helvetica Chimica Acta*, **59**, 1099–111.

Kodama, M. & Kimura, E. (1976). *Journal of The Chemical Society, Dalton Transactions*, 2341–5.

Lehn, J.-M. & Sauvage, J.-P. (1975). *Journal of the American Chemical Society*, **97**, 6700–7.

Michaux, G. & Reisse, J. (1982). *Journal of the American Chemical Society*, **104**, 6895–9.

Micheloni, M. & Paoletti, P. (1980). *Inorganica Chimica Acta*, **43**, 109–12

Micheloni, M., Paoletti, P. & Sabatini, A. (1983). *Journal of The Chemical Society, Dalton Transactions*, 1189–91.

Micheloni, M., Paoletti, P., Siegfried-Hertli, L. & Kaden, T. A. (1985). *Journal of The Chemical Society, Dalton Transactions*, 1169–72.

Petersen, C. J. & Frensdorff, H. K. (1972). *Angewandte Chemie, International Edition*, **11**, 16–25.

Pett, V. B., Diaddario, L. L., Dockal, E. R., Corfield, P. W., Ceccarelli, C., Glick, M. D., Ochrymowycz, L. A. & Rorabacher, D. B. (1983). *Inorganic Chemistry*, **22**, 3661–70.

Schmidt, E., Tremillon, J.-M., Kintzinger, J.-P. & Popov, A. I. (1983). *Journal of the American Chemical Society*, **105**, 7563–6.

Sokol, L. S. W. L., Ochrymowycz, L. A. & Rorabacher, D. B. (1981). *Inorganic Chemistry*, **20**, 3189–95.

Thom, V. J., Hosken, G. D. & Hancock, R. D. (1985), *Inorganic Chemistry*, **24**, 3378–81.

Thom, V. J., Shaikjee, M. S. & Hancock, R. D. (1986). *Inorganic Chemistry*, **25**, 2992–3000.

Thom, V. J. & Hancock, R. D. (1985). *Journal of The Chemical Society, Dalton Transactions*, 1877–80.

Chapter 7

Anderegg, G., Ekstrom, A., Lindoy, L. F. & Smith, R. J. (1980). *Journal of the American Chemical Society*, **102**, 2670–4.

Bemtgen, J. M., Springer, M. E., Loyola, V. M., Wilkins, R. G. & Taylor, R. W. (1984). *Inorganic Chemistry*, **23**, 3348–53.

Billo, E. J. (1984). *Inorganic Chemistry*, **23**, 236–8.

Cabbiness, D. K. & Margerum, D. W. (1970). *Journal of the American Chemical Society*, **92**, 2151–3.

Chock, P. B. (1972). *Proceedings of the National Academy of Science, USA,* **69**, 1939–42.

Cox, B. G., Garcia-Rosas, J. & Schneider, H. (1981). *Journal of the American Chemical Society*, **103**, 1054–9.

Cox, B. G., Schneider, H. & Stroka, J. (1978). *Journal of the American Chemical Society*, **100**, 4746–9.

Cox, B. G., van Truong, N. & Schneider, H. (1984). *Journal of the American Chemical Society*, **106**, 1273–80.

Diaddario, L. L., Zimmer, L. L., Jones, T. E., Sokol, L. S. W. L., Cruz, R. B., Yee, E. L., Ochrymowycz, L. A. & Rorabacher, D. B. (1979). *Journal of the American Chemical Society*, **101**, 3511–20.

Drumhiller, J. A., Montavon, F., Lehn, J. M. & Taylor, R. W. (1986). *Inorganic Chemistry*, **25**, 3751–7 and references therein.

Ekstrom, A., Leong, A. J., Lindoy, L. F., Rodger, A., Harrison, B. A. & Tregloan, P. A. (1983). *Inorganic Chemistry*, **22**, 1404–7.

Ekstrom, A., Lindoy, L. F., Lip, H. C., Smith, R. J., Goodwin, H. J., McPartlin, M. & Tasker, P. A. (1979). *Journal of The Chemical Society, Dalton Transactions*, 1027–31.

Ekstrom, A., Lindoy, L. F. & Smith, R. J. (1980). *Inorganic Chemistry*, **19**, 724–7.

Graham, P. G. & Weatherburn, D. C. (1981). *Australian Journal of Chemistry*, **34**, 291–300.

Grant, C. & Hambright, P. (1969). *Journal of the American Chemical Society*, **91**, 4195–8.

Hambright, P. (1971). *Coordination Chemistry Reviews*, **6**, 247–68.

Hay, R. W., Bembi, R., McLaren, F. & Moodie, W. T. 1984). *Inorganica Chimica Acta*, **85**, 23–31.

Hay, R. W., Bembi, R., Moodie, W. T. & Norman, P. R. (1982). *Journal of The Chemical Society, Dalton Transactions*, 2131–6.

Hay, R. W., Jeragh, B., Lincoln, S. F. & Searle, G. H. (1978). *Inorganic & Nuclear Chemistry Letters*, **14**, 435–40.

Hay, R. W. & Norman, P. R. (1980). *Inorganica Chimica Acta*, **45**, L139–41.

Hertli, L. & Kaden, T. A. (1981). *Helvetica Chimica Acta*, **64**, 33–7.

Kodama, M. & Kimura, E. (1977). *Journal of The Chemical Society, Dalton Transactions*, 2269–76.

Leugger, A. P., Hertli, L. & Kaden, T. A. (1978). *Helvetica Chimica Acta*, **61**, 2296–300.

Liesegang, G. W. (1981). *Journal of the American Chemical Society*, **103**, 953–5.

Liesegang, G. W., Farrow, M. M., Purdie, N. & Eyring, E. M. (1976). *Journal of the American Chemical Society*, **98**, 6905–8.

Liesegang, G. W., Farrow, M. M., Vazquez, F. A., Purdie, N. & Eyring, E. M. (1977). *Journal of the American Chemical Society*, **99**, 3240–3.

Lin, J. D. & Popov, A. I. (1981). *Journal of the American Chemical Society*, **103**, 3773–7.

Lin, C. T., Rorabacher, D. B., Cayley, G. R. & Margerum, D. W. (1975). *Inorganic Chemistry*, **14**, 919–25 and references therein.

Lincoln, S. F., Horn, E., Snow, M. R., Hambley, T. W., Brereton, I. M. & Spotswood, T. M. (1986). *Journal of The Chemical Society, Dalton Transactions*, 1075–80, and references therein.

Lindoy, L. F. & Smith, R. J. (1981). *Inorganic Chemistry*, **20**, 1314–6.

Loyola, V. M., Pizer, R. & Wilkins, R. G. (1977), *Journal of the American Chemical Society*, **99**, 7185–8.

Kallianou, C. S. & Kaden, T. A. (1979). *Helvetica Chimica Acta*, **62**, 2562–8.

Klaehn, D.-D., Paulus, H., Grewe, R. & Elias, H. (1984). *Inorganic Chemistry*, **23**, 483–90.

Margerum, D. W., Cayley, G. R., Weatherburn, D. C. & Pagenkopf, G. K. (1978). *Advances in Chemistry Series*, **174**, 1–220.

Margerum, D. W., Rorabacher, D. B. & Clarke, J. F. G. (1963). *Inorganic Chemistry*, **2**, 667–77.

Maynard, K. J., Irish, D. E., Eyring, E. M. & Petrucci, S. (1984). *The Journal of Physical Chemistry*, **88**, 729–36.

Melson, G. A. & Wilkins, R. G. (1963). *Journal of The Chemical Society*, 2662–72.

Murphy, L. J. & Zompa, L. J. (1979). *Inorganic Chemistry*, **18**, 3278–81.

Rao, V. H. & Krishnan, V. (1985). *Inorganic Chemistry*, **24**, 3538–41, and references therein.

Riedo, T. J. & Kaden, T. A. (1979). *Helvetica Chimica Acta*, **62**, 1089–96.

Rodriguez, L. J., Liesegang, G. W., Farrow, M. M., Purdie, N. & Eyring, E. M. (1978). *The Journal of Physical Chemistry*, **82**, 647–50.

Rodriguez, L. J., Liesegang, G. W., White, R. D., Farrow, M. M., Purdie, N. & Eyring, E. M. (1977). *The Journal of Physical Chemistry*, **81**, 2118–22.

Schmidt, E. & Popov, A. I. (1983). *Journal of the American Chemical Society*, **105**, 1873–8.

Shamim, A. & Hambright, P. (1983). *Inorganic Chemistry*, **22**, 694–6, and references therein.

Shchori, E., Jagur-Grodzinski, J. & Shporer, M. (1973). *Journal of the American Chemical Society*, **95**, 3842–6.

Vitiello, J. D. & Billo, E. J. (1980). *Inorganic Chemistry*, **19**, 3477–81.

Wilkins, R. G. (1974). *The Study of Kinetics and Mechanism of Reactions of Transition Metal Complexes*. Allyn and Bacon, Inc., Boston.

Chapter 8

Ansell, C. W. G., Lewis, J., Ramsden, J. N. & Schroder, M. (1983). *Polyhedron*, **2**, 489–91.

Bailey, C. L., Bereman, R. D., Rillema, D. P. & Nowak, R. (1984). *Inorganic Chemistry*, **23**, 3956–60.

Barefield, E. K. & Mocella, M. T. (1973). *Inorganic Chemistry*, **12**, 2829–32.

Bond, A. M., Lawrance, G. A., Lay, P. A. & Sargeson, A. M. (1983). *Inorganic Chemistry*, **22**, 2010–21.

Boucher, H. A., Lawrance, G. A., Lay, P. A., Sargeson, A. M., Bond, A. M., Sangster, D. F. & Sullivan, J. C. (1983). *Journal of the American Chemical Society*, **105**, 4652–61.

Busch, D. H. (1978). *Accounts of Chemical Research*, 392–400 and references therein.

Buttafava, A., Fabbrizzi, L., Perotti, A., Poggi, A. & Seghi, B. (1984). *Inorganic Chemistry*, **23**, 3917–22.

Carnieri, N., Harriman, A., Porter, G. & Kalyanasundaram, K. (1982). *Journal of The Chemical Society, Dalton Transactions*, 1231–8 and references therein.

Chang, D., Malinski, T., Ulman, A. & Kadish, K. M. (1984). *Inorganic Chemistry*, **23**, 817–24.

Chen, Y.-W. D. & Bard, A. J. (1984). *Inorganic Chemistry*, **23**, 2175–81 and references therein.

Curtis, N. F. (1968). *Coordination Chemistry Reviews*, **3**, 3–47.

Curtis, N. F. (1971). *Journal of The Chemical Society, A*, 2834–8.

Curtis, N. F. (1974). *Journal of The Chemical Society, Dalton Transactions*, 347–59 and references therein.

Curtis, N. F. & Cook, D. F. (1967). *Journal of The Chemical Society, Chemical Communications*, 962–3.

Dabrowiak, J. C., Lovecchio, F. V., Goedken, V. L. and Busch, D. H. (1972). *Journal of the American Chemical Society*, **94**, 5502–4.

Fabbrizzi, L. (1979). *Inorganica Chimica Acta*, **36**, L391–3.

Fabbrizzi, L. (1985). *Comments in Inorganic Chemistry*, **4**, 33–54.

Fabbrizzi, L., Lari, A., Poggi, A. & Seghi, B. (1982). *Inorganic Chemistry*, **21**, 2083–5.

Freiberg, M., Meyerstein, D. & Yamamoto, Y. (1982). *Journal of The Chemical Society, Dalton Transactions*, 1137–41.

Goedken, V. L. & Busch, D. H. (1972). *Journal of the American Chemical Society*, **94**, 7355–63.

Groves, J. T. and Gilbert, J. A. (1986). *Inorganic Chemistry*, **25**, 125–7 and references therein.

Haines, R. I. & McAuley, A. (1982). *Coordination Chemistry Reviews*, **39**, 77–119.

Harrowfield, J. MacB., Herlt, A. J., Lay, P. A., Sargeson, A. M., Bond, A. M., Mulac, W. A. & Sullivan, J. C. (1983). *Journal of the American Chemical Society*, **105**, 5503–5.

Hill, C. L. & Schardt, B. C. (1980). *Journal of the American Chemical Society*, **102**, 6374–5.

Hipp, C. J., Lindoy, L. F. and Busch, D. H. (1972). *Inorganic Chemistry*, **11**, 1988–94.

Hung, Y., Martin, L. Y., Jackels, S. C., Tait, A. M. & Busch, D. H. (1977). *Journal of the American Chemical Society*, **99**, 4029–38.

Jubran, N., Cohen, H., Koresh, Y. & Meyerstein, D. (1984). *Journal of The Chemical Society, Chemical Communications*, 1683–4.

Kelly, S. L. & Kadish, K. M. (1982), *Inorganic Chemistry*, **21**, 3631–9 and references therein.

Kestner, M. O. & Allred, A. L. (1972). *Journal of the American Chemical Society*, **94**, 7189–90.

Lever, A. B. P., Minor, P. C. and Wilshire, J. P. (1981). *Inorganic Chemistry*, **20**, 2550–3.

Lovecchio, F. V., Gore, E. S. & Busch, D. H. (1974). *Journal of the American Chemical Society*, **96**, 3109–18.

Olson, D. C. & Vasilevskis, J. (1969). *Inorganic Chemistry*, **8**, 1611–21.

Olson, D. C. & Vasilevskis, J. (1971). *Inorganic Chemistry*, **10**, 463–70.

Palmer, J. M., Papaconstantinou, E. & Endicott, J. F. (1969). *Inorganic Chemistry*, **8**, 1516–23.

Patterson, G. S. & Holm, R. H. (1975). *Bioinorganic Chemistry*, **4**, 257–75.

Rakowski, M. C. & Busch, D. H. (1973). *Journal of the American Chemical Society*, **97**, 2570–1.

Rorabacher, D. B., Martin, M. J., Koenigbauer, M. J., Malik, M., Schroeder, R. R., Endicott, J. F. & Ochrymowycz, L. A. (1983). In *Copper Coordination Chemistry: Biochemical and Inorganic Perspectives*, ed. K. D. Karlin & J. Zubieta, pp. 167–202. Adenine, Guilderland, New York.

Tait, A. M., Lovecchio, F. V. & Busch, D. H. (1977). *Inorganic Chemistry*, **16**, 2206–12.

Truex, T. J. & Holm, R. H. (1972). *Journal of the American Chemical Society*, **94**, 4529–38.

Walker, D. D. & Taube, H. (1981), *Inorganic Chemistry*, **20**, 2828–34.

Zeigerson, E., Bar, I., Bernstein, J., Kirschenbaum, L. J. & Mayerstein, D. (1982). *Inorganic Chemistry*, **21**, 73–80.

Chapter 9

Almog, J., Baldwin, J. E., Dyer, R. L., Huff, J. and Wilkerson, C. J. (1974). *Journal of the American Chemical Society*, **96**, 5600–1.

Almog, J., Baldwin, J. E. & Huff, J. (1975). *Journal of the American Chemical Society*, **97**, 227–8.

Almog, J., Baldwin, J. E., Dyer, R. L. & Peters, M. (1975) *Journal of the American Chemical Society*, **97**, 226–7.

Anderson, D. L., Weschler, C. J. & Basolo, F. (1974). *Journal of the American Chemical Society*, **96**, 5599–600.

Baldwin, J. E. & Huff, J. (1973). *Journal of the American Chemical Society*, **95**, 5757.

Baldwin, J. E. & Perlmutter, P. (1984). *Topics in Current Chemistry*, **121**, 181–220.

Basolo, F., Hoffman, B. M. & Ibers, J. A. (1975). *Accounts of Chemical Research*, **8**, 384–92.

Behr, J. P., Kirch, M. & Lehn, J.-M. (1985). *Journal of the American Chemical Society*, **107**, 241–6.

Bresciani-Pahor, N., Forcolin, M., Marzilli, L. G., Randaccio, L., Summers, M. F. & Toscano, P. J. (1985). *Coordination Chemistry Reviews*, **63**, 1–125.

Brinigar, W. S. & Chang, C. K. (1974). *Journal of the American Chemical Society*, **96**, 5595–7.

Chang, C. K., Koo, M. S. & Ward, B. (1982). *Journal of The Chemical Society, Chemical Communications*, 716–19.

Chang, C. K. & Traylor, T. G. (1973). *Journal of the American Chemical Society*, **95**, 8477–9 and references therein.

Chottard, G., Schappacher, M., Ricard, L. & Weiss, R. (1984). *Inorganic Chemistry*, **23**, 4557–61.

Collman, J. P. (1977). *Accounts of Chemical Research*, **10**, 265–72.

Collman, J. P., Gagne, R. R., Gray, H. B. & Hare, J. W. (1974). *Journal of the American Chemical Society*, **96**, 6522–4.

Collman, J. P., Gagne, R. R., Reed, C. A., Robinson, W. T. & Rodley, G. A. (1974). *Proceedings of the National Academy of Science, USA*, **71**, 1326–9.

Collman, J. P., Halbert, T. R. & Suslick, K. S. (1980). In *Metal Ion Activation of Dioxygen*, ed. T. G. Spiro. Wiley, New York.

Dickerson, R. E. & Timkovich, R. (1975). In *The Enzymes*, ed. P. D. Boyer. Vol. XI. Academic Press, New York.

Diebler, H., Eigen, M., Ilgenfritz, G., Maass, G. & Winkler, R. (1969). *Pure and Applied Chemistry*, **20**, 93–115.

Dobler, M., Dunitz, J. D. & Kilbourn, B. T. (1969). *Helvetica Chimica Acta*, **52**, 2573–83.

Dobler, M., Dunitz, J. D. & Krajewski, J. (1969). *Journal of Molecular Biology*, **42**, 603–6.

Dolphin, D., (ed.), (1978–9). *The Porphyrins*, Vols. I–VII. Academic Press. New York.

Farmery, K. & Busch, D. H. (1970). *Journal of The Chemical Society, Chemical Communications*, 1091.

Fenton, D. E. (1977). *Chemical Society Reviews*, 325–43.

Fermi, G. (1975). *Journal of Molecular Biology*, **97**, 237–56.

Goldsby, K. A., Beato, B. D. & Busch, D. H. (1986). *Inorganic Chemistry*, **25**, 2342–7.

Groves, J. T. & Gilbert, J. A. (1986). *Inorganic Chemistry*, **25**, 125–7.

Groves, J. T., Kruper, W. J. & Haushalter, R. C. (1980). *Journal of the American Chemical Society*, **102**, 6375–7.

Groves, J. T., Nemo, T. E. & Myers, R. S. (1979). *Journal of the American Chemical Society*, **101**, 1032–3.

Groves, J. T., Watanabe, Y. & McMurray, T. J. (1983). *Journal of the American Chemical Society*, **105**, 4489–90.

Gunter, M. J., Mander, L. N., McLaughlin, G. M., Murray, K. S., Berry, K. J., Clark, P. E. & Buckingham, D. A. (1980). *Journal of the American Chemical Society*, **102**, 1470–3.

Hay, R. W. (1984). *Bio-Inorganic Chemistry*, Ellis Horwood, Chichester, UK.

Hilgenfeld, R. & Saenger, W. (1982). *Topics in Current Chemistry*, **101**, 1–82.

Izatt, S. R., Hawkins, R. T., Christensen, J. J. & Izatt, R. M. (1985). *Journal of the American Chemical Society*, **107**, 63–6.

Izatt, R. M., Lamb, J. D., Hawkins, R. T., Brown, T. R., Izatt, S. R. & Christensen, J. J. (1983). *Journal of the American Chemical Society*, **105**, 1782–5.

James, B. R. (1978). In *The Porphyrins* ed. D. Dolphin. Academic Press. New York and London.

Lamb, J. D., Christensen, J. J., Izatt, S. R., Bedke, K., Astin, M. S. & Izatt, R. M. (1980). *Journal of the American Chemical Society*, **102**, 3399–403.

Lamb, J. D., Izatt, R. M. & Christensen, J. J. (1981). *Stability Constants of Cation-Macrocycle Complexes and Their Effect on Facilitated-Transport Rates*, Ch 2 in *Progress in Macrocyclic Chemistry*, Volume 2, ed. R. M. Izatt & J. J. Christensen. Wiley-Interscience, New York.

Lauger, P. (1972). *Science*, **178**, 24–30.

Livorness, J. & Smith, T. D. (1982). *Structure and Bonding*, **48**, 1–44.

McLendon, G. & Martell, A. E. (1976). *Coordination Chemical Reviews*, **19**, 1–39.

Malmstrom, B. G. (1980) in *Metal Ion Activation of Dioxygen*, ed. T. G. Spiro. Wiley, New York.

Neupert-Laves, K. & Dobler, M. (1975). *Helvetica Chimica Acta*, **58**, 432–42.

Ochiai, E. & Busch, D. H. (1968). *Journal of The Chemical Society, Chemical Communications*, 905–6.

Painter, G. R. & Pressman, B. C. (1982). 'Dynamic Aspects of Ionophore Mediated Membrane Transport', in *Host Guest Complex Chemistry* II, ed. F. Vogtle. Springer-Verlag, Berlin.

Parker, W. O., Zangrando, E., Bresciani-Pahor, N., Randaccio, L. & Marzilli, L. G. (1986). *Inorganic Chemistry*, **25**, 3489–97.

Pauling, L. (1964). *Nature*, **203**, 182–3.

Perutz, M. F. (1971). *New Scientist and Science Journal*, 676–9.

Perutz, M. F. (1978). *Scientific American*, **239**, 92–125.

Phillips, S. E. V. (1980). *Journal of Molecular Biology*, **142**, 531–54.

Phillips, S. E. V. & Schoenborn, B. P. (1981). *Nature*, **292**, 81–2.

Racker, E. (1979). *Accounts of Chemical Research*, **12**, 338–44.

Schrauzer, G. N. (1971). *Advances in Chemistry Series*, **100**, 1–20.

Shaanan, B. (1982). *Nature*, **296**, 683–4.

Smegal, J. A. & Hill, C. L. (1983). *Journal of the American Chemical Society*, **105**, 2920–2.

Smith, K. M. Ed. (1975). *Porphyrins and Metalloporphyrins*. Elsevier, New York.

Sykes, A. G. (ed.) (1982). *Advances in Inorganic and Bioinorganic Mechanism*, Vol. 1, pp. 121–78. Academic Press, New York.

Takano, T. (1977). *Journal of Molecular Biology*, **110**, 569–84.

Taylor, R. W., Kauffman, R. F. & Pfeiffer, D. R. (1982). 'Cation Complexation and Transport by Carboxylic Acid Ionophores' in *Polyether Antibiotics*, ed. J. W. Westley, Vol. 1, 103–84.

Ten Eyck, L. F. & Arnone, A. (1976). *Journal of Molecular Biology*, **100**, 3–11.

Van Gelder, B. F. & Beinert, H. (1969). *Biochemistry Biophysics Acta*, **189**, 1–24.

Wagner, G. C. & Kassner, R. J. (1974). *Journal of the American Chemical Society*, **96**, 5593–5.

Wang, J. H. (1958). *Journal of the American Chemical Society*, **80**, 3168–9.

Index

Printed in the United States
By Bookmasters